SECOND EDITION

Secrets to Success in Sport & Play

A Practical Guide to Skill Development

Marianne Torbert, PhD

Leonard Gordon Institute for Human Development Through Play
Temple University

Human Kinetics

Library of Congress Cataloging-in-Publication Data

Torbert, Marianne.
Secrets to success in sport & play : a practical guide to skill development /
Marianne Torbert. -- 2nd ed.
p. cm.
Includes bibliographical references and index.
ISBN-13: 978-0-7360-9029-2 (soft cover)
ISBN-10: 0-7360-9029-0 (soft cover)
1. Physical education and training. 2. Human mechanics. 3. Motor learning. I. Title.
GV341.T64 2011
613.7--dc22
2010039250

ISBN-10: 0-7360-9029-0 (print)
ISBN-13: 978-0-7360-9029-2 (print)

This book is a revised edition of *Secrets to Success in Sport & Play: A Guide for Players of All Ages,* published in 1982 by Prentice-Hall, Inc.

The Web addresses cited in this text were current as of September 2010, unless otherwise noted.

Acquisitions Editor: Scott Wikgren; **Developmental Editor:** Melissa Feld; **Assistant Editor:** Rachel Brito; **Copyeditor:** Patrick Connolly; **Indexer:** Dan Connolly; **Permission Manager:** Dalene Reeder; **Graphic Designer:** Robert Reuther; **Graphic Artists:** Robert Reuther and Denise Lowry; **Cover Designer:** Keith Blomberg; **Art Manager:** Kelly Hendren; **Associate Art Manager:** Alan L. Wilborn; **Illustrator:** © Human Kinetics; **Printer:** Versa Press

Printed in the United States of America 10 9 8 7 6 5 4 3 2 1

The paper in this book is certified under a sustainable forestry program.

Human Kinetics
Web site: www.HumanKinetics.com

United States: Human Kinetics, P.O. Box 5076, Champaign, IL 61825-5076
800-747-4457
e-mail: humank@hkusa.com

Canada: Human Kinetics, 475 Devonshire Road Unit 100, Windsor, ON N8Y 2L5
800-465-7301 (in Canada only)
e-mail: info@hkcanada.com

Europe: Human Kinetics, 107 Bradford Road, Stanningley, Leeds LS28 6AT, United Kingdom
+44 (0) 113 255 5665
e-mail: hk@hkeurope.com

Australia: Human Kinetics, 57A Price Avenue, Lower Mitcham, South Australia 5062
08 8372 0999
e-mail: info@hkaustralia.com

New Zealand: Human Kinetics, P.O. Box 80, Torrens Park, South Australia 5062
0800 222 062
e-mail: info@hknewzealand.com

E5062

To the many who love the joy and celebration of sport and play as players, teachers, parents, and spectators.

It is my hope that this book will allow each reader to enjoy additional involvement, growth, and skill.

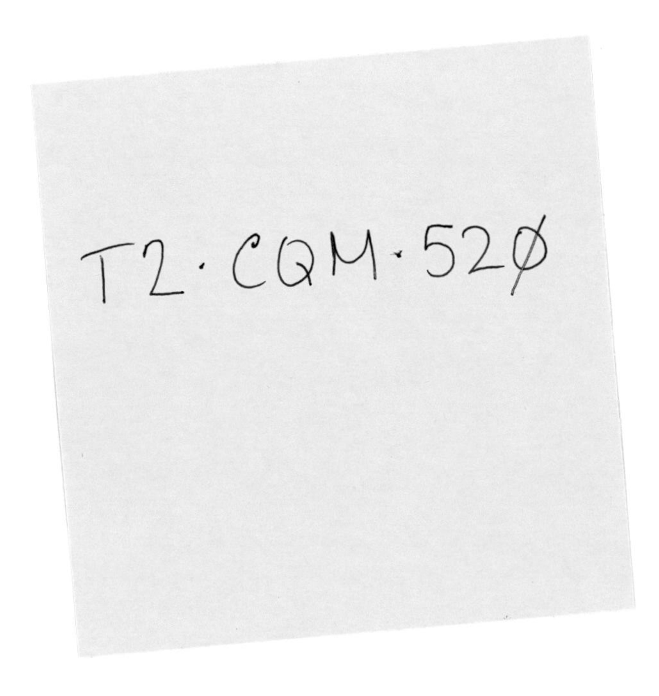

CONTENTS

FOREWORD

Secrets to Success in Sport & Play is the second book of what I hope will be a series by Marianne Torbert. Countless other instructional books deal with the mechanics of sport and games. Unfortunately, many of them contain misconceptions based on either the folklore of sport or the unsubstantiated opinions of popular sports figures. An understanding of the mechanics of skilled human movement requires a sound insight into the concepts of Newtonian physics and some familiarity with the anatomical structure of the body. Thus, the information that the author conveys must be scientifically unambiguous and at times complex. This book appeals to a diverse field. Therefore, the most formidable task in writing this book was to take each concept and present it in a manner that would be readily comprehensible to nonspecialized readers.

Clearly, Dr. Torbert has succeeded in ensuring that the significance of the concepts was not lost through oversimplification. The book is dedicated to the many who love the joy of sport and play, and you will sense that the author shares that love. Her informal writing style and her personal anecdotes are both refreshing and applicable to those in the field. Most important, *Secrets to Success in Sport & Play* promotes learning through participation rather than by reading alone. These characteristics are the hallmarks of a real educator.

Marianne Torbert's book is for players of all ages and for parents, teachers, and coaches. You most likely will search through this book in the hope of finding the secrets to success in sport and play, and you will indeed uncover those secrets. But in the search, you will also find some of the joys and the magic of science. Perhaps this is Dr. Torbert's greatest gift to her readers.

Peter R. Francis
Past chairperson
Kinesiology Academy of the American Alliance for Health,
Physical Education, Recreation and Dance

PREFACE

Play has an expanding and unlimited potential to contribute to personal growth. Increasing one's understanding of any aspect of play will increase the personal challenge to be found within it. This challenge will enable people to produce and increase physical skills. At the same time, the challenge can contribute to the quality of people's lives by adding to the thrill of playing, reducing daily accidents, opening doors to social interaction with others, and perhaps increasing participants' personal confidence.

This book can help you do all of this. I have selected several principles of movement and attempted to incorporate them in a way that will clarify how you can apply them and make them work for you. Here are some things you should keep in mind:

1. No one is destined to be unskilled.
2. Understanding the keys to the mechanics of sport and play can greatly increase your success; it can also reduce the time normally required to improve your skills.
3. All individuals can improve their balance. By increasing your balance, you will also improve your agility, power, and accuracy.
4. Knowledge of spins can improve your performance in many sports, including basketball, tennis, bowling, soccer, and golf.
5. Understanding some of the basic mechanics of movement can help you begin to become an outstanding coach, teacher, or player.

Over the years, I have spent a lot of time teaching people to play. While doing so, I frequently found that participants as well as some coaches really had no idea about WHY things happen. There seemed to be a great deal of interest, however, whenever I attempted to clarify and simplify the mechanical principles of movement that had so excited me when I discovered them in various texts. As I fumbled in my attempts to explain certain principles, I found I was learning a great deal myself. A desire to simplify led me to seek help in biomechanics books, and what I found excited me even further.

Although I had been a nationally ranked participant, I was now discovering secrets to success in sport and play that I had been unaware of. It was challenging to try out new possibilities and conquer old unsolved problems. I found myself beginning to be able to transfer my new understandings from one activity to another—and that's when the contents of this book began to evolve. The book has gone through the normal labor pains as decisions of inclusion, omission, and format had to be made. My usual but unbelievable luck gave me help from Dr. Peter Francis of the University of Oregon, chairperson of the Academy of

Kinesiology of the American Alliance for Health, Physical Education, Recreation and Dance. His assistance enabled me to maintain scientific honesty in my attempts to simplify and explain materials that frequently involve a somewhat sophisticated understanding of math, engineering, and physics.

Secrets to Success in Sport & Play was written to assist both those who are seeking personal help and those who wish to help others in this task. The format, which is meant to be participatory, was selected to maximize your sense of involvement. You will find suggestions, observations, experiments, and applications that you might like to try as you read. Each of these has been expanded in the new edition. For this edition of *Secrets to Success in Sport & Play,* several new features have been added. Each chapter has been updated with many additional sport applications for each mechanical principle. For each chapter, a set of true or false review questions now appears at the end of the chapter; appendix C provides the answers to these questions along with explanations regarding why a question is true or false. You'll also find a new chapter (10) that describes numerous games that have proven to be helpful in assisting players to learn and apply the mechanical principles found in this book. This chapter includes a matrix that allows you to match games to principles at a glance. Also, a new appendix contains information on how you can make your own equipment.

Play may be the key to open many doors to lifelong growth and development. Understanding seems to lead to further understanding, and gradually you are actually playing with the secrets to success in sport and play. For some of you, the enjoyment of learning can become play in itself as you help yourself and others move well. My hope is that the reading of this book and your participation in the activities described will be a pleasurable experience that encourages you to have a lifelong involvement in play. Robert W. White, a Harvard psychologist, once said, "Play may be fun, but it is also a serious business in childhood. During these hours the child steadily builds up his (or her) competence in dealing with the environment." I would raise only one question: Why should this be limited to childhood? I think you'll find that the games and activities in this book can be played by all ages.

My personal reason for writing this book is my desire to share the joy I have known in successful participation in sport and play—and the joy of overcoming the difficulties within a partnership of the physical and the mental. I believe you will find that the excitement of participation increases as you begin to discover the secrets to success in sport and play.

True or False Review Questions

Information about each question is found on the page number provided within the parentheses following the question.

REMEMBER THAT A PARTIALLY FALSE QUESTION IS CONSIDERED FALSE.

Answers to all questions are found in appendix C.

1. An understanding of the mechanical principles of movement allows a person to have a better understanding of why things happen—and in turn increases the person's ability to decide how those things might be done better (page xi).
2. Writing this book was easy because all the author had to do was simplify the math, engineering, and physics involved in the mechanical principles of movement (page xi).
3. *Secrets to Success in Sport & Play* was written only for coaches (page xii).
4. Understanding the keys to the mechanics of sport and play can reduce daily accidents and the time normally required to improve skills (page xi).

ACKNOWLEDGMENTS

Thanks to all of the following:

- Peter R. Francis, professor emeritus of San Diego State University, for his patient and careful assistance in trying to keep me scientifically honest. Any errors in judgment are strictly mine and are probably the result of my bullheadedness.
- Marion Broer, whose books turned me on to the mechanical principles of movement.
- My students (especially Gene White, Brian Barrett, Steve Palis, Lisa Hand, Joan Wood, Jody Kelly, Ron Quinn, and Nancy Stout), who helped me to understand Kahlil Gibran's statement: "Thought is a bird of space, that in a cage of words may indeed unfold its wings but cannot fly."
- Lynne B. Schneider, who has patiently put up with me during the stressful times and has helped me to clarify my thoughts.
- Morgan Beatty, who has taught me some very special things about the magic of play.
- Doug Parise, whose knowledge of movement, artistic talent, and personal patience and perseverance made the initial illustrations for this book possible.
- The College of Health, Physical Education, Recreation, and Dance; the College of Education; the College of Health Professions; and the department of kinesiology that made my tenure at Temple University so pleasurable and have allowed me the privilege of being the director of the Leonard Gordon Institute for Human Development Through Play (www.temple.edu/leonardgordoninstitute).
- Scott Wikgren of Human Kinetics, who enthusiastically supported the publishing of this book through every step of the process.
- Rainer Martens, who was wise in his awareness that the study of human movement needed a strong publishing company.
- The Cleveland Women's Physical Education Association, which gave its first scholarship to a scared, five-foot-two, 100-pounder who had a dream of becoming a physical education teacher. That dream came true.
- And last, but certainly not least, my brother, Raymond Richard Rothhaas, who taught me so many things that have meant so much to me through the years.

PART I

Approaching Movement Analysis

CHAPTER 1

Teaching and Learning

Learning involves input, processing, experiencing, evaluating, and trying again. This can be facilitated by the assistance of a knowledgeable teacher, coach, parent, or player who is willing to take an analytical problem-solving approach that involves using the mechanical principles of movement.

This chapter covers two approaches to analysis. One approach—the specific objective method—involves looking at your objective, or the specific skill that you want to develop. The second approach—the correcting errors method—involves dealing with errors that are occurring.

Specific Objective Method

Determine your *objective*. What do you want to accomplish? Begin to list the mechanical principles that might help you achieve your objective. Perhaps you want to hit the tennis ball over the net consistently. This is certainly an important goal. You could detail and study all the mechanical principles that might be involved in accomplishing this task (as done in appendix A); however, a better strategy would be to analyze the specific need that you are going to deal with first and to direct your efforts toward this goal. For example, if you're not making contact with the ball, perhaps the problem lies in your inability to visually evaluate a moving object (see chapter 9). If you contact the ball but it just doesn't go far enough, you have a force problem and could use the mechanical principles in chapter 4 as a checklist. If you find your force improving but now the ball is going wild, then the mechanical principles related to direction and accuracy (described in chapter 8) should be helpful. Try to enjoy the process of puzzling out what mechanical principles are related to your objective. The chapters in this book should give you an excellent repertoire of possibilities.

If you are trying to correct an error that is causing a problem, you should consider a different approach: the correcting errors method.

Correcting Errors Method

The correcting errors approach assumes that a player is having specific problems, so the focus becomes the errors and the need to correct these errors. Correcting a single error will often correct many other errors as well. The most relevant error that may be affecting success can be tackled through analysis.

To identify the most relevant error, you may need to become a mental detective. Here is a method that you might like to try:

- State the problem as you see it.
- Begin to zero in on the key to the solution by using the "curious child" technique (*why, why, why?*). Ask why the stated problem occurred. For example, you may determine the following: "The ball went too high instead of going low and directly over the net." Then try to determine what may have caused this problem to occur (e.g., "I hit under the ball"). When you have this answer, you should ask why *this* problem occurred (e.g., "I didn't swing level" or "My racket was tilted slightly upward"). The key is to find the first problem or error that occurred, because this error is probably the cause of all the subsequent errors. Identifying the first error makes it easier to determine the mechanical principles that will be the most helpful. Look through the chapters in this book and jot down possibilities. When you correct the earliest occurring error, the errors that followed will tend to correct themselves. If you only correct the later errors (which are really just effects or results of the initial error), the initial error will not be corrected, and thus the primary problem will still exist.

The key is to find the first problem or error that occurred, because this error is probably the cause of all the subsequent errors.

Attempt to put your solution into practice. Be patient. Remember, correcting an error takes time. It may be more difficult than other forms of learning because you may also be working on breaking habits.

Practice your observational skills. You can learn a great deal from watching both good and bad players. Can you determine why the former succeed and why the latter fail? If you need activities to observe, try the games in chapter 10.

Initially, the study of the biomechanical principles of movement may take some real effort. But with time you will begin to see that "the principles of balance, force production, motion and leverage are identical regardless of the activity" (Broer and Zernicke 1979, p. 29). The purpose of each (movement pattern) causes some adjustments, but the basic mechanics remain the same (p. 13). An understanding of the principles or procedures underlying the initial task will result in greater transfer to a different activity (Oxendine 1968, p. 97) and greater depth of understanding on your part.

The principles of balance, force production, motion and leverage are identical regardless of the activity (Broer and Zernicke 1979, p. 29).

In working with either the specific objective method or the correcting errors method, you should be aware that most movements have three phases:

1. *Preparatory phase:* backswing, stabilization, weight transfer away from the final direction of action, stretching of the muscles to be involved in the action phase

2. *Action phase:* motion or effort that follows the preparatory phase and precedes the final follow-through phase

3. *Follow-through phase:* completion of a movement after the action phase; absorption of force; reaching out after a hit, throw, or kick. This may lead into a new preparatory stage (e.g., when catching a throw, you rock backward to absorb the force).

An understanding of the principles or procedures underlying the initial task will result in greater transfer to a different activity (Oxendine 1968, p. 97) and greater depth of understanding on your part.

Any limitations in a previous phase will negatively affect the phases that follow.

Encouraging the Learner

Here are some tips that you should keep in mind when helping players learn skills:

- Remember that feelings are important to learning and changing habits.
- Encourage players to try not to invest their identity in instant success. Mistakes are also a vital part of learning.
- Try to recognize improvement—even when it may seem minimal.
- Use a visual model. A visual model gives learners an idea of what they are supposed to be doing. This can be a demonstration, a picture, or observation of a live or recorded performance. I have noted that learners tend to focus on the results rather than concentrate on the three phases of the motion; therefore, I no longer demonstrate by hitting a ball *over* the net or shooting a ball *into* the basket. Learners retain more of the relevant aspects of the demonstration when the result becomes irrelevant. So I demonstrate only the motion that I want the learners to focus on and retain. No ball or goal is involved.
- Remember that young children may not be the best listeners. Try to help them understand and begin to use the mechanical principles of movement through *experiencing*. Games and movement activities can be selected that will enable them to practice specific foundation skills that incorporate the mechanical principles. These foundation skills (balance, visual tracking, absorbing force, changing directions, stopping and starting, spatial [space] awareness, and reading movement) are vital because they underlie many activity skills. (See the games in chapter 10.)
- Try to have a thorough understanding of the mechanical principles and the visual evaluation skills involved so that your help can be specific and keyed to a particular skill or problem. This understanding will help you see relationships, make wise choices and decisions, and increase the possibility of a transfer of learning from one activity or situation to another.
- Avoid giving too much information at once. Solve only one problem at a time. Focus on what is most relevant and will give the players the most to build on. Problem areas will be discussed throughout the chapters that follow.

Try to have a thorough understanding of the mechanical principles and the visual evaluation skills involved so that your help can be specific and keyed to a particular skill or problem. This understanding will help you see relationships, make wise choices and decisions, and increase the possibility of a transfer of learning from one activity or situation to another.

- Try to develop special key words and phrases that seem to be helpful and are understood by as many players as possible. The "gorilla" technique discussed in chapter 3 (page 24) is an example.

- Be patient. The tension created by stress is both an emotional and a physical deterrent to learning. Players need time and lots of repetition in order to compute what works; to get the "feeling" for moving (kinesthetic sense); to recognize flight and rebound patterns; to learn to know what to attend to; to time their moves; and to let extraneous, ineffective, or overflow movements be extinguished.

- Try to determine "progressions" for growth. Consider reducing the complexity or the number of problems to be solved at any one time. Try to simplify, perhaps breaking the skill down or using a lead-up activity. Try to reduce the number of things in motion. Use a tee, tether, or trough (figure 1.1). Use balloons, beach balls, or other slow-moving objects. Toss slowly and accurately to the beginner. Increase the size of the object or striking implements without increasing the weight. And try to make initial experiences consistent, simple, and successful.

- Help the players establish obtainable goals, and help them see and feel their successes. Positive feelings about growth and improvement increase motivation and challenge, and they reduce threat, another deterrent to learning.

- Make sure that each participant has *many* trials, is challenged at her level, and has some success.

- Create an environment that supports trying and allows for the errors that normally occur during learning.

- Try to reduce the fears of failure and of injury. Sometimes equipment can be effectively modified. Be creative. See pages 6-8 and appendix B.

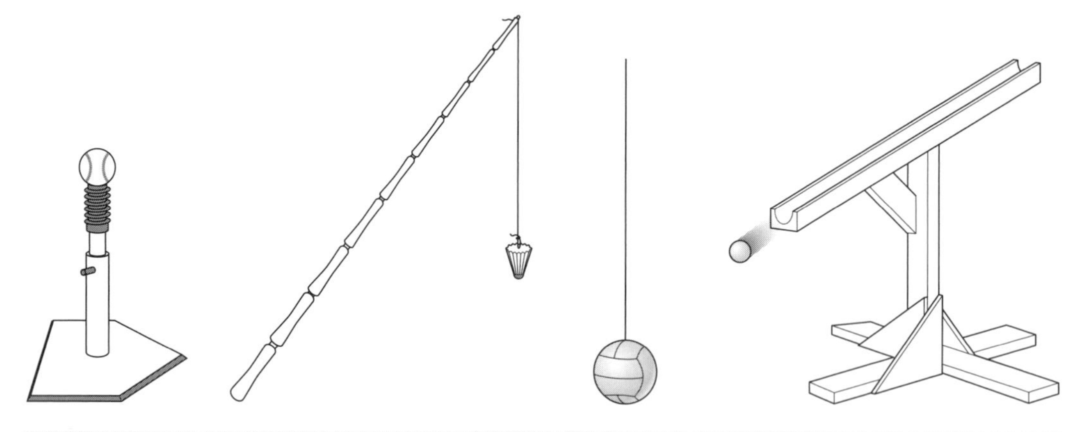

FIGURE 1.1 Using a tee, tether, or trough reduces the number of things in motion and increases consistency, allowing the player to focus on one factor at a time.

Make sure that each participant has *many* trials, is challenged at her level, and has some success.

- If necessary, allow players to compensate until they gain the ability to be forceful, to be accurate, and to move well. For instance, players may need to send the ball high to get it over the net (in volleyball), use a lighter ball or bat, or lower the basket. These techniques are helpful during the process of learning and can be modified as abilities develop and improve.

- Try to avoid correcting a player in front of others because this could be embarrassing. You might do a role change in which you let the participants (individual or group) find an error that *you* demonstrate. This allows you to avoid pointing out who made the error, and it encourages players to become good observers.

- If necessary, provide players with creative "helps" that enable them to sense specific body positions or develop specific foundation skills. Be inventive. Have fun creating new helps. Remember that we do not all learn in the same way or through the same methods. Here are some examples of helps:

    1. Use a rope with a knot in the far end of it to help players get the feeling of the full reach of an overhead throw or tennis serve. Players should swing the rope through the movement pattern of the throw or tennis serve.
    2. Play catch with plastic gallon jug scoops to learn about "giving" to absorb force (see Partner Scoop Play in chapter 10).
    3. To help a beginner get the feeling of a full backswing, have the player reach back and touch a fence, your hand, or a hanging object.
    4. To help players get the feeling of a level swing (not scooping or dropping the bat or racket), have the players swing bats or rackets along a tennis net or rope.
    5. Set up two or three tees (batting or golf), each with a ball on it. Have players hit through the balls to get the feeling of a level swing, flattening the swinging arc, and keeping their eye on the ball (figure 1.2; also see Batting [or Golf] Tee Challenge in chapter 10). If you don't have three adjustable batting tees, or if you want to learn how to make a three-tee golf device, refer to appendix B for information on how to make your own equipment.

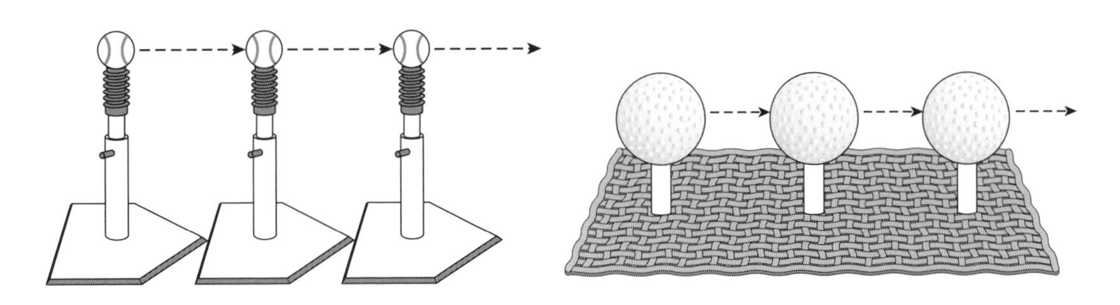

FIGURE 1.2 Multiple tees and multiple turns allow players to get the feel for the swing while encouraging them to keep their eye on the ball.

6. To help a tennis player who is learning to serve, suspend a paper plate at the proper height and position in relation to the individual. Then ask the player to ball toss gently and accurately to just touch the plate (figure 1.3).
7. Have volleyball players "set" the ball through a basketball hoop. This encourages good technique while providing a built-in motivation—that is, a clear and immediate measurement of the degree of success (figure 1.4).

- Help players avoid fighting old habit patterns. Try starting with what is *unfamiliar* but related. For example, have tennis players do the backhand before trying to break all the bad habits that go with the self-taught forehand. Have a child do the backward roll before the forward roll. (Actually, the loss of balance that initiates the roll is accomplished more readily in the backward roll.) Later, you can relate what was learned to the familiar and hope for positive transfer.
- Be aware of and sensitive to the problems of those who may do things differently: the nearsighted individual, the heavyset player, the awkward adolescent who is dealing with a changing body, the individual who feels uncomfortable attempting a new skill, the tense individual, the left-handed player, and so on. For example, instead of using terms such as *left* or *right,* consider terms that can apply to both left- and right-handed players, such as *other, opposite, net foot,* or *racket hand.*
- Encourage thinking. Attempt to build a learning progression that will allow for independent learning as players develop in skill and understanding. One possible approach over time might be as follows:

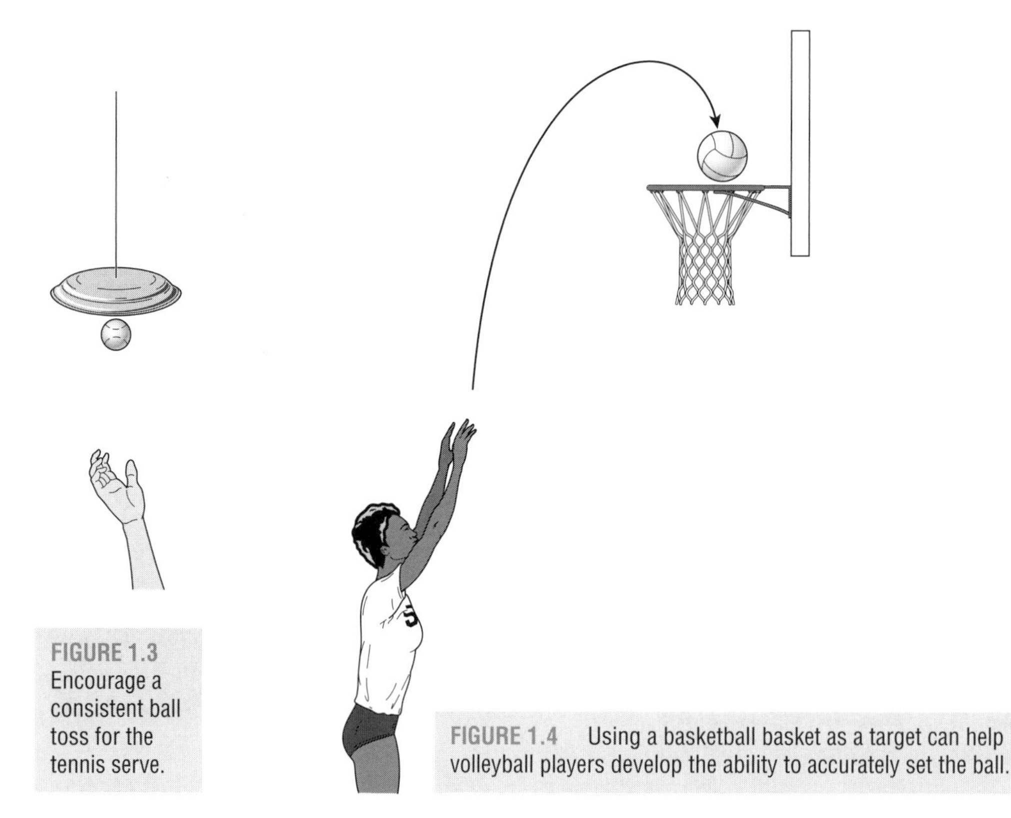

FIGURE 1.3 Encourage a consistent ball toss for the tennis serve.

FIGURE 1.4 Using a basketball basket as a target can help volleyball players develop the ability to accurately set the ball.

1. Tell beginners how to do something.
2. Later, ask participants what you told them.
3. Tell participants why something is done a certain way (based on the mechanical principles of movement). Stay simple.
4. Later, ask players why something is done a certain way. (You may be interested in finding out just what the players thought you said.)
5. Show beginners similarities between various skills and situations. Encourage generalization and transfer of learning; for instance, this can be done with the overhand throw, the badminton smash, and the tennis serve.
6. Guide players to see the transfer of similarities between various skills, such as bowling and the underhand throw.
7. Ask players to tell you about similarities between old and new activities.
8. Ask players to relate mechanical principles to the objective of a particular known skill.
9. Ask participants to relate principles to the objective of an unknown or new skill.
10. Ask players to make decisions about how to do something based on their growing understanding.

For various reasons (e.g., fear, initial failure, a rapidly changing body, size, unreal expectations, vision problems, lack of experience at a very basic level, or individual uniqueness), many players need and can grow from a situation that allows for multiple repetitions in a safe and comfortable environment. This may require sensitivity, patience, and creative selection or modification on the part of the coach, parent, or teacher.

This type of help can be of tremendous value during any part of an individual's life and may contribute very positively to the pleasure that he enjoys while playing. I had an interesting experience that taught me that although you do not always directly witness this effect, it may still happen. I was in an elevator in a large office building in Columbus, Ohio. A tall, lanky young man got on and, looking right at me, said, "Hello, Ms. Torbert, I'll bet you don't know who I am." His face seemed familiar, but my recall allowed me little more than a vague hunch that this young man's name was Rick. He *was* Rick; the last time I had seen him he was two-thirds as tall and 10 to 12 years younger. He had been the awkward, persevering kid who had taken free tennis lessons with me years before. To my surprise, he had gone on to do quite well. He knew and had played with nationally ranked tennis players. He had come a long way—much further than I would have predicted a decade earlier while watching him struggle to get to and hit the ball. Back then, he would struggle trying to deal with his body and the complexity of a bouncing ball. Rick thanked me for the initial start he had received from those early lessons. But as he left and the door closed behind him, I realized that he had just given me a bigger gift. He had let me know that the help we give, sometimes wondering if it has been of any value, may reap unseen results. And that potential end is so worth the effort.

True or False Review Questions

Information about each question is found on the page number provided within the parentheses following the question.

REMEMBER THAT A PARTIALLY FALSE QUESTION IS CONSIDERED FALSE.

Answers to the questions and additional information are found in appendix C.

1. In the specific objective method, the term *specific objective* refers to the specific sport that you want to deal with (page 3).

2. Each chapter in *Secrets to Success in Sport & Play* deals with a particular sport (thumb through the book).

3. The correcting errors method assumes that a player has become consistent enough to be having a specific problem (page 4).

4. The correcting errors method involves asking "why, why, why?" in order to get to the initial error that is probably the cause of all later errors (page 4).

5. You can learn a great deal from observing both good and bad players (page 4).

6. The author has found that learners retain more of the relevant aspects of a demonstration when the result of the action is successful (page 5).

7. Games and fun movement activities can be selected to help players improve basic skills, such as balance, visual tracking, and absorbing force (page 5).

8. The principles of balance, force production, motion, and leverage are identical regardless of the activity (page 4).

9. Basic skills such as absorbing force, spatial (space) awareness, changing directions, and stopping and starting are sport specific and need to be learned for each specific sport (pages 4-5).

References

Broer, M.R., and R.F. Zernicke. 1979. *Efficiency of Human Movement.* Philadelphia: W.B. Saunders Company.

Oxendine, J.B. 1968. *Psychology of Motor Learning.* New York: Appleton-Century-Crofts.

PART II

Moving Yourself

CHAPTER 2

Balance

Balance is the foundation from which we initiate all movement. Without balance, many tasks become relatively difficult or even impossible. We cannot effectively develop force nor can we hope for accuracy, consistency, or coordination without good balance. Observe a skillful athlete and it becomes evident that the capability to lose and regain balance is the key to agility (the ability to change directions rapidly), maneuverability, and all other efficient uses of the body. Improving balance increases the degree of control and adaptability that athletes have in relation to their movements.

For people who do not move well or are considered clumsy, balance should be an important area of focus and practice. Many people have embarrassing moments, have more than the average number of injuries, and fail unnecessarily because they have not sufficiently developed their balance. In at least three periods of life, people may need to learn or redevelop some aspects of this physical skill—toddlerhood, adolescence (when the body may be in a rapid state of change), and as a senior citizen (when some of the physical faculties that we have relied on for balance may begin to diminish, including sight, muscular endurance, strength, flexibility, and the mechanism in the semicircular canals in the ears).

Because balance can be improved at any age and is of lifetime importance to each person's social, emotional, and physical well-being, people need to participate frequently in fun, safe activities involving balance. We often overlook the opportunity to include this type of activity as part of prepractice warm-ups, in family play, at picnics, or as a fun conclusion to a practice session. All players, both children and adults, need a diet of fun, noneliminating types of games, lots of practice on specific skills, and opportunities for evaluation and improvement. (See the activities in chapter 10, especially the balance activities.) These are a vital part of the process of developing and maintaining balance skills. Activities involving balance can be played by people of any age and any level, from the beginner (or the player who has balance difficulties) to the varsity-level player.

We cannot effectively develop force nor can we hope for accuracy, consistency, or coordination without good balance.

Principles of Balance

Although balance can be improved through experience and practice, people also need to understand the three mechanical principles that govern balance. People who understand these principles can help themselves and others by analyzing the causes of specific balance difficulties and by making helpful suggestions or decisions that can greatly reduce the amount of trial and error required to increase balance ability. This in turn reduces the injury potential, increases the success ratio, and allows participants to experience success more rapidly. The three important principles of balance are as follows:

1. Keep your weight centered over your base of support.

   c of g
   ———
   b of s

2. Increase the size of your foundation (base of support)—for example, spread your feet.

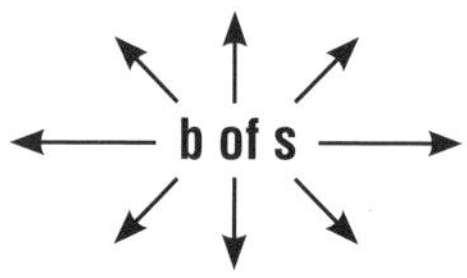

   ← **b of s** →

3. Lower your weight (center of gravity)—for example, bend your knees.

   ↓ **c of g** ↓

The concept of the center of gravity is important for those who wish to understand the principles of balance and the principles related to the human being as a projectile (see chapter 7). A person's center of gravity is the point around which the body can be balanced. Although it is found roughly behind the umbilicus (belly button) in the pelvic area, the center of gravity will vary with body build and will change with each new position of the body. Thus, if we shift our weight (for instance, raise an arm, putting more body weight above the waist), the center of gravity must also shift in this direction to continue to balance the weight. If we move a leg backward and upward, as in the preparatory backswing for a kick, the center of gravity will also shift backward and upward within the body. These changes frequently require a shift of the body position so that the center of gravity can remain over the base of support for stability.

As the center of gravity moves nearer the outer border of the base of support, relative instability occurs. When the center of gravity moves beyond the base of support, the body is no longer in balance (stability), and the person will have to shift his position to bring the center of gravity back over the base of support; otherwise, instability, falling, or movement will occur (figure 2.1).

Because balance can be improved at any age and is of lifetime importance to each person's social, emotional, and physical well-being, people need to participate frequently in fun, safe activities involving balance.

There are actually two kinds of balance, both of which are important in different situations. The first is called static balance, or minimal movement balance (such as in a

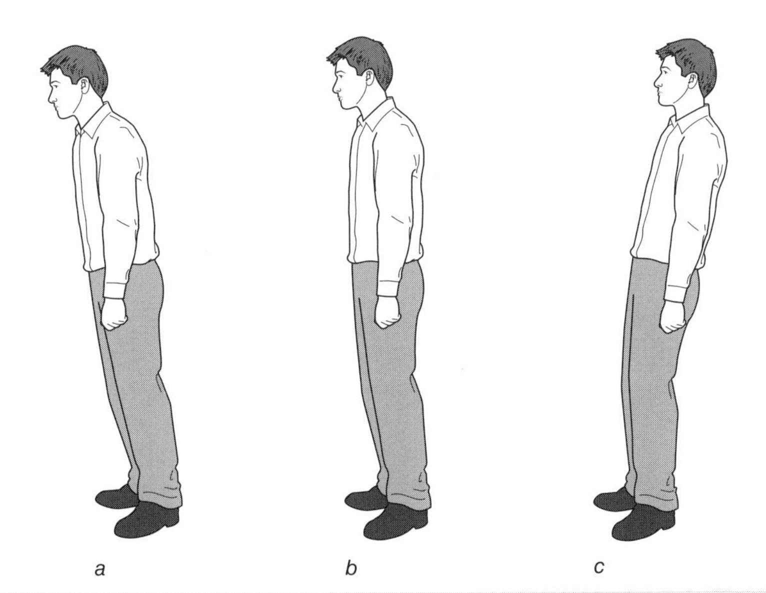

FIGURE 2.1 As the center of gravity moves beyond the base of support, balance becomes precarious.

headstand). For this type of balance, a person attempts to make only slight shifts to keep the center of gravity over the base of support and to maintain balance in a relatively still (stable) state. The second type is dynamic balance, or moving balance. This type tends to involve a continuous process of gaining, deliberately losing, and then regaining one's balance, all in a well-controlled moving state, such as when walking or running. Both of these kinds of balance abilities are vital to good play.

People can also get moving by deliberately breaking the rules governing stability. Because moving balance (instability) is the opposite of stability (balance that involves attempting to maintain a still position), you can get moving simply by putting your center of gravity outside (beyond) your base of support in any direction you desire to move. When you do this, you will automatically be on the move. For instance, when performing a start in sprint racing, the runner takes advantage of a loss of balance (figure 2.2). A raised center of gravity and a reduced (or small)

FIGURE 2.2 Sprinters can use a loss of balance to their advantage.

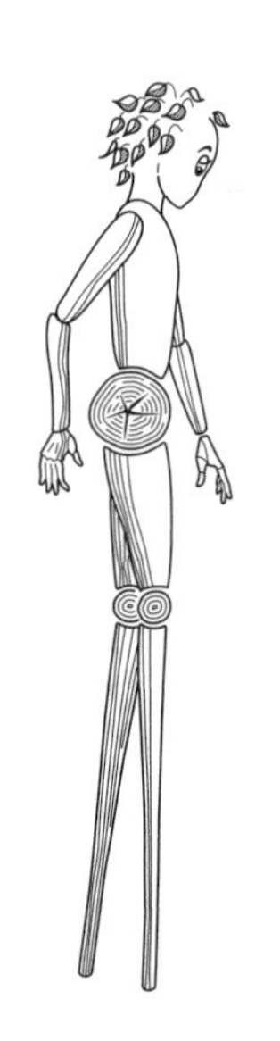

FIGURE 2.3 A high center of gravity and a small base of support make balance more difficult and challenging.

base of support could also contribute to initiating this movement (figure 2.3).

When losing balance, you will tend to start moving somewhat rapidly and without effort, because the pull of gravity will help do the job for you. This is why some people have described running as a series of falls in which runners continually lose and catch their balance (figure 2.4). Can you find situations in which losing balance contributes to a quick start? Can you see how and why people must shift their center of gravity to maintain balance when stopping or changing directions?

In some circumstances, an uncontrolled loss of balance could lead to a serious injury. A person can avoid this loss of balance by incorporating one or more of the principles of balance. Can you find specific situations in which you could help another person or yourself avoid a loss of balance by employing one or more of the three balance principles?

How to Improve Your Balance

Everyone can improve their balance and save themselves from possible injuries and embarrassment. The following tips may help:

- Understand and apply the mechanical principles of balance. Consider these in varied and real situations.
- Explore and practice a variety of static and dynamic balance positions. Remember that games and other movement activities can be an excellent source of balance practice (see Skates, Frantic Ball, Freeze, Rag or Rug Hockey, and Super Sox in chapter 10). Also remember that people can learn at any age to adapt more effectively to messages received from visual input, the semicircular canals in the ears, pressures on the body, and the proprioceptors (a sensory system found within the muscles and tendons).
- Learn to relax. Being able to relax improves your reception of and responses to the messages sent by body mechanisms. People who need to improve their balance must learn to relax, and the environment in which they move must initially be as emotionally supportive as possible. They can then learn to remain relaxed under progressively more stressful conditions.

FIGURE 2.4 Is running actually a continual loss of balance?

- Develop sufficient muscular strength and endurance in the muscles of the abdomen and lower limbs. This enables a person to make necessary balance adjustments.

Everyone can improve their balance and save themselves from possible injuries and embarrassment.

If you want to improve balance while standing, landing, moving, or stopping suddenly, you should keep the following tips in mind:

1. Bend the knees instead of keeping them straight. (Can you see the important role that the knees play in both lowering the center of gravity and allowing the body weight to be adjusted over the base of support?) Assuming a bent-knee stride position with a lean into the oncoming force will allow you to absorb the force gradually in a rocking motion, shifting your weight toward the oncoming force and then rocking backward to absorb this force to maintain your balance. Rocking backward may also be useful as a preparatory movement for your next action (see Super Sox in chapter 10).

2. Increase the size of the base of support in the direction in which you may need to adjust your balance. A base of support that is wide but not in the direction of the oncoming force can actually reduce your adaptive ability to maintain your balance.

3. When stopping suddenly, lean away from—not toward—the direction you were moving. This keeps the center of gravity over the base of support.

4. When landing or stopping suddenly, keep your head up and look forward. Don't look at the ground by dropping your head forward. (Can you see how dropping your head forward would shift your weight forward, possibly decreasing your stability?)

Observations

Observe sport events. When players must change direction suddenly, what do they do to maintain their moving (dynamic) balance? Note how they lean. How does this affect their center of gravity? What about the size of their base of support? How do they position their feet? How much of their foot is in contact with the ground? You will understand the function of this practice and its relationship to maintaining balance after you have read chapter 4 (specifically, the information on absorbing force). Can friction on a rough surface affect balance by helping a player absorb force? Can a lack of friction on a smooth surface make it more difficult to maintain balance? How do the different kinds of footwear worn in different sports, on different surfaces, and under varying conditions help in maintaining balance?

Note how the degree of stability is determined by how directly the weight is centered over the base of support. Can you see the adaptive function that the knees and other joints play in this situation? Note how the skilled performers shift their center of gravity toward an oncoming force to allow for good balance adjustment as the center of gravity is pushed backward. This is similar to leaning into the wind when you are walking on a very windy day.

Various playing positions and situations may require varying degrees of stability and instability. Can you see examples of this in football or wrestling? If a player was uncertain about what would happen next, what position (in relation to the center of gravity and the base of support) might prove most effective? Can you relate this to boxing, fencing, or tag? Does good balance seem to have any effect on a player's ability to make rapid adjustments?

Consider what a football blocker does to unbalance the opponent. Does a blocker lift the opponent's center of gravity? Does he attempt to push the opponent's center of gravity beyond the base of support? Does he attempt to reduce the opponent's base of support? Can you see how each one of these contributes to causing the opponent to lose balance? Can you see why pushing without lifting would not be as effective?

A wrestler might push and lift in a similar way. Can you see why poor posture might decrease good balance? How does the pole used in tightrope walking aid the performer's balance? Consider the adaptability factor in relation to shifting the center of gravity for lateral (side-to-side) balance. Can you see how the non-racket arm in tennis is important for maintaining good balance? Who might have more stability, a tall or short player? A player with even weight distribution or a player with a large upper trunk?

In some instances, a lack of flexibility could affect a person's potential for balance adjustment. Can you determine situations in which a senior citizen's balance could be affected by a lack of flexibility? Looking at figure 5.6 (in chapter 5), can you see how an athlete whose iliopsoas muscles are tight could be pulled into a swayback position? Could this have an effect on balance? Note both the position of the center of gravity and the adaptability factor.

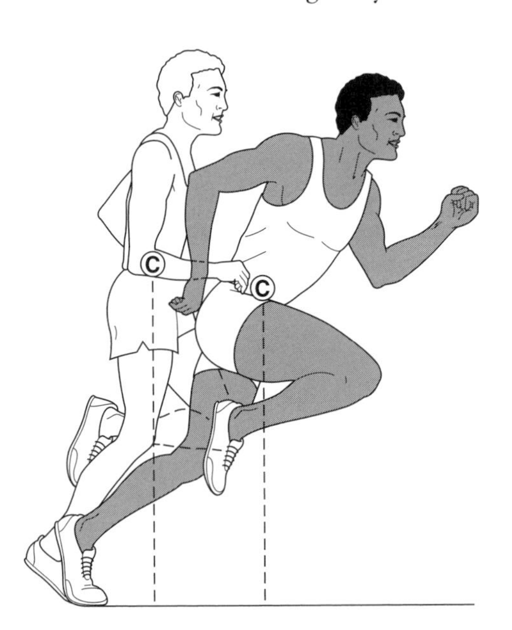

FIGURE 2.5 The position of the center of gravity differs in jogging and running. Does this affect speed?

Observe a person who appears clumsy. Can you determine if any of the principles of balance are being violated? What suggestions might you make?

Can you see how tension or tension-producing stress could affect balance? Can you see why running is faster than jogging? Where is the center of gravity in relation to the base of support during running and jogging (figure 2.5)? Can you see how the direction of lean could initiate the forward or backward movement in tumbling and gymnastics (figure 2.6)?

Activities

Some individuals may have a sense of queasiness or awkwardness when putting their bodies into new positions or positions not experienced for some time. This is a normal response and will diminish as involvement continues. Games that challenge one's balance can be a source of improvement. Refer to the game matrix on page 100 in chapter 10. Encouraging relaxation also helps. Here are some activities that can help improve balance:

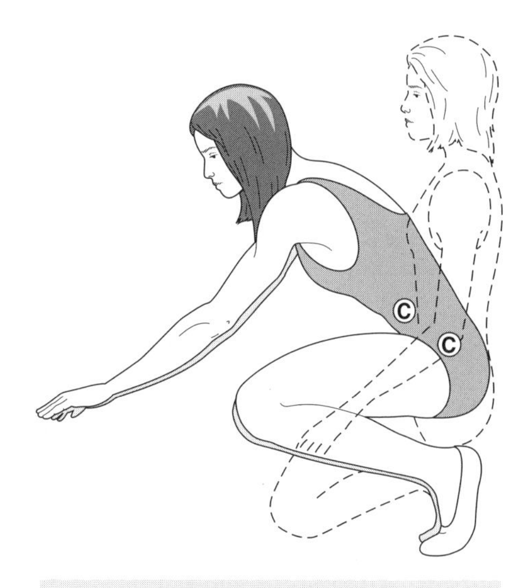

FIGURE 2.6 Gymnasts use the center of gravity and the base of support to cause movement to occur.

- Practice moving in slow motion. Challenge yourself with difficult balance positions. (See Freeze on page 107 in chapter 10.)
- Try to balance on one foot. Now close your eyes. Our eyes are important in helping us to balance. If you are going to stay balanced with your eyes closed, you must learn to use other sources of information.
- Place your feet directly below your shoulders. Now have someone push you gently from the front or behind. Next, place your feet in a forward and backward stride position and ask someone to push again. Compare the two. If you were receiving a forceful pass in basketball, which position would you prefer?
- Perform hand wrestling. With hands gripped as shown in figure 2.7, a good base of support, and knees bent to lower the center of gravity, each participant attempts to force the other off balance. You might be able to equalize the competition between two people of different abilities by having the winner of each trial go on to a more difficult balance position to challenge herself on the next trial (e.g., a long, narrow base of support with one foot directly in line with the other or knees straight and stiff). Also see Super Sox in chapter 10 (page 119).
- Do a basketball defense shuffle or shadow practice. Take a guarding position. Attempt to shadow (follow exactly) a moving player (figure 2.8). Note how you use your feet, ankles, knees, and other body parts to help you change directions and maintain balance.
- Run and come to a sudden stop. How do you maintain your balance? Observe someone who tends to lose balance when stopping. Can you determine which mechanical principles he needs to work on?
- Try the following games from chapter 10 for improving balance: Frantic Ball, Freeze, Pass-a-Puck, Rag or Rug Hockey, Skates, and Super Sox. These games also help people develop the leg and abdominal strength and endurance needed to hold and adjust their balance.

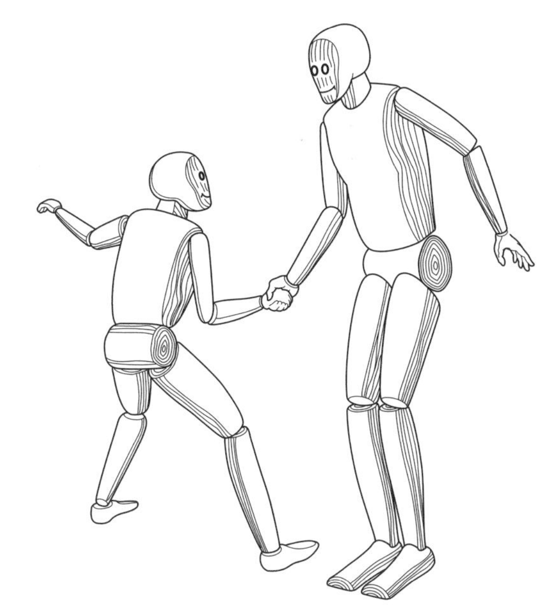

FIGURE 2.7 Hand wrestling can be a fun way to improve balance.

FIGURE 2.8 The basketball defense shuffle or shadowing is a good way to improve one's ability to change directions quickly while continuing to maintain a balance position in preparation for the next move.

I'd like to tell you a story about a kindergartener, Scott, who was quite large for his age. He was anything but aggressive, but he was known to hurt other children by bumping into them or falling on them during playtime. This kept the other kindergarteners in fear and away from him. It limited his social contacts and in turn his social growth.

At the end of the school year, Scott's teacher thought he should repeat kindergarten so that he could improve his social skills. This seemed like an illogical remedy because he would be even larger in relation to the new kindergarten children, who would be a year younger than him.

The boy's mother was in a class of mine at the university. She brought the problem to class with her, told her story, and asked for help. The students asked her a great many questions. They tried to trace the problem backward to the initial problem of social immaturity based on limited social experience and based on other children being afraid to play with Scott. They also considered why Scott had bumped into or fallen on other children when he really just wanted to play with them. His mother said that he frequently had bumps and bruises from falls and seemed clumsier than other children. But she also said that she had also had a lot of bumps and bruises growing up and that she just assumed it was part of childhood. The class decided that because of Scott's size, and because of his lack of opportunity to play, the real problem might be poor balance.

Frantic Ball, Freeze, Pass-a-Puck, Rag or Rug Hockey, Skates, and Super Sox help people develop the leg and abdominal strength and endurance needed to hold and adjust their balance.

That summer Scott's family put him on carpet "skates" (pages 117-118 in chapter 10) and played lots of games with him. Carpet skates were chosen because they would help Scott work on improving his balance. By the end of the summer, Scott had better balance (evidenced by fewer bruises) and had a wonderful repertoire of fun games to share with his classmates.

I wish this story ended with Scott being allowed to go on to first grade with his classmates. But in reality, Scott's social skills were still limited, and he was held back to repeat kindergarten.

But it was a good year for Scott. The new kindergarteners were not afraid of him because he no longer bumped into or fell on the other children. Actually, he became a popular playmate because he not only played well, but he had all these neat games to share.

At the end of Scott's second kindergarten year, he was promoted to first grade; his written report read "social maturity = excellent; a class leader who loves to play and help others."

True or False Review Questions

Information about each question is found on the page number provided within the parentheses following the question.

REMEMBER THAT A PARTIALLY FALSE QUESTION IS CONSIDERED FALSE.

Answers to the questions and additional information are found in appendix C.

1. Balance is the foundation from which we initiate all movement. We cannot effectively develop consistent force, accuracy, or coordination without good balance (page 13).
2. Most sport skills programs for adolescents and senior citizens stress balance development. This is important because these are two periods of life when balance may have to be redeveloped (page 13).
3. When the body weight is shifted (for instance, when the person raises an arm or leg), the center of gravity of the body shifts in the opposite direction to continue to balance the weight (page 14).
4. A loss of balance is a way to get started quickly (pages 15-16).
5. Bending the knees is a way to incorporate two of the three principles of balance (pages 14, 17).
6. To absorb an oncoming force, you should widen your base of support (page 17).
7. A wrestler or football blocker can best cause his opponent to lose his balance by hitting him hard through the center of gravity (page 18).
8. Activities and games that involve a bent-knee position, kinesthetic awareness, abdominal development, the building of leg strength and endurance, or the adjustment of the center of gravity over a changing base of support will help to develop balance (pages 18-19).

CHAPTER 3

Initiating Movement

Do you need to be fast, forceful, or both? The answer to this question will help you determine what you might do to get started.

Every time we begin to move or change directions, we must first deal with *inertia.* Inertia is sometimes thought of as a state of stillness, but this definition is incomplete because inertia can be a still or moving state. Inertia is actually any present state of being, and overcoming inertia involves a change in that state and the resistance to that change. It could even involve a change in either speed (faster or slower) or direction. Change of any kind requires a force to overcome inertia (our present state). In this chapter, we deal only with ways to help overcome inertia to initiate a move.

Many people do not give a great deal of consideration to this initial aspect of moving, but it is a vital component and may very well determine whether success or failure will follow. It can affect the amount of force that can be generated and how quickly we can accomplish a given task. Tasks usually require force, speed, or varying combinations of these components. So we must clearly delineate what we are attempting to do and must determine whether force or speed is most important to the specific task. Understanding that various options are available and that some of these options can more effectively meet our needs than others, we must select the way we will initiate our move based on the task at hand. Decisions are task or goal specific, and people need to know where they are going and how soon they must be there before they choose how to get started.

Getting Ready to Move

- *Are you physiologically ready?* Are you warmed up for the task? Warming the body increases the blood flow and reduces the sluggishness of body tissues and fluids.

Many people do not give a great deal of consideration to this initial aspect of moving, but it is a vital component and may very well determine whether success or failure will follow.

We can compare the fluidlike state within the body tissues to the oil in your car. If you start your car when it is cold, the oil is thicker. Once the car is running, the oil warms and becomes thinner. Now the engine can work less to accomplish the same task, because the counteracting forces have been reduced, increasing the freedom of movement. When you play, be sure you warm up and remain warm. This will allow you to move more quickly and forcefully and to avoid injuries that could occur if you were not sufficiently internally warmed.

- *Are you anatomically ready?* You save time if you already have your involved joints bent or unlocked and if the muscles to be used are on a partial stretch. Bending the joints in preparation for the move can put these muscles on a partial stretch.

Watch anyone begin to move. The person must initiate the movement by bending the involved joints. Find a tennis, softball, or volleyball player who stands upright with knees straight and you can assume that this person will start to move more slowly than necessary. A muscle that is stretched has a greater number of muscle fibers ready for involvement. A muscle fiber that is already shortened cannot contract to cause movement to occur. Keeping the knees slightly bent does require more effort, but well-toned muscles can maintain this position without fatigue. In fact, this very act will help to maintain the needed muscle tone and keep the muscles warm.

Two of my friends who coach Little League, Bob Gallagher and Ed Dunn, use a clever technique to accomplish this position of readiness. They tell their players to be like gorillas. It works! Not only is the word *gorilla* a quick, positive reminder, but the word and the technique that goes with it can be easily remembered.

Bent knees may be the key to increased success in many activities.

The bent-knee ready position is also important because by bending the knees, the person automatically lowers the center of gravity for better balance. This position also allows for a shift of weight (center of gravity in relation to the base of support) to adjust more rapidly to the unexpected. Wrestlers, fencers, or tag players who rarely get tagged must be ready at all times for whatever happens (figure 3.1).

FIGURE 3.1 Are you ready to move? What will help? Initiating movement quickly is important in many sports.

Observe their body position. Bent knees may be the key to increased success in many activities.

• *A full and rapid backswing can stimulate the stretch reflex, which assists in rapidly overcoming inertia during initial acceleration.*

Sensory receptors, called muscle spindles, are stimulated when skeletal muscles are forcefully stretched, thus activating the stretch reflex. These receptors are part of a reflex mechanism that "fires" a message to the spinal cord. Because reflex impulses travel only to the spinal cord and not on to the higher nervous system, the responding impulse for the muscle to contract is very rapid. Thus, while the stretch reflex mechanism protects the muscles against being overstretched and torn, it can also help us to overcome inertia very rapidly in the initial period of acceleration. This becomes important in force development because momentum is the product of mass times velocity. In addition, players must not confuse the use of the stretch reflex to rapidly overcome inertia (thus allowing more force to be developed over the available distance) with the rapid completion of an action, which will be discussed later. Evoking the stretch reflex requires *additional* time during the preparation phase and thus would not be used in a situation that involves the rapid completion of a task.

Evoking the stretch reflex requires *additional* time during the preparation phase and thus would not be used in a situation that involves the rapid completion of a task.

A muscle that is not well conditioned could be damaged if it is rapidly or forcibly stretched. For this reason, you should maintain adequate muscle development and should be warmed up before deliberately involving the stretch reflex.

Watch the preparatory phase of a pitcher, a field goal kicker, or a weightlifter (see figure 3.2). Can you see how these athletes use the stretch reflex to their advantage? Golfers and tennis players also use it to gain greater force when they want to send the ball a long distance.

The value of stretching a muscle before using it will be discussed further in chapter 4.

FIGURE 3.2 A full and rapid backswing can stimulate the stretch reflex and assist in a rapid overcoming of inertia to increase force.

• *In some situations, you can use the pull of gravity to initiate movement quickly.*

As noted in chapter 2, when a person's center of gravity is beyond the person's base of support, a condition of instability or movement occurs. Sprinters, football line players, and gymnasts frequently use this source of force to help them initiate movement quickly (figure 3.3).

Have you ever observed a child who rocks backward before getting started in the direction she wishes to go? This rocking action will increase the force that can be applied to the rear foot, but it may take time, causing the child to get a slow start. Some activities require

FIGURE 3.3 A deliberate loss of balance can create a quick start.

the player to initiate movement rapidly in order to be successful. For example, a child who gets a slow start may be tagged, may be slow to reach base, or may be hit by the oncoming dodgeball. At times, a player needs to sacrifice a specific force technique for one that allows her to get started rapidly and thus complete the task more rapidly. If this child could learn to use gravity to initiate her movements rapidly rather than use a rear-foot push, success in these activities might be enhanced.

As you move, try to find ways to use the pull of gravity to your advantage in overcoming inertia. Can you find at least one way you can use gravity to help you move more effectively?

- *You can shorten levers to get there faster with less force.* Lever length and force will be discussed further in chapter 4.

Force will sometimes become secondary, less important, or even unnecessary. Starting rapidly, completing a task in a given amount of time, being ready for the next move, repeating a motion several times as rapidly as possible, or reducing or absorbing force may become the primary objective.

If a quick move or change is needed and force is not a primary factor, shortening the involved levers may help. (See the section in chapter 4 on lever length.) The arms, legs, and trunk are actually all levers. Note the difference between a short-distance runner and a marathoner. Observe how much more rapidly the sprinter's legs come around. This requires more energy expenditure, but it allows for increased speed when that is your primary concern. To accomplish this, the legs (levers) are shortened by bending the knees more. The arms are flexed to allow the rotation time of the arms and legs to be synchronized.

A short throw from the infield to first base, a "flick" in badminton, a block volley at the net in tennis, and a bunt in softball or baseball all require little force but need a rapid response for success. This can be accomplished by shortening the lever (in these instances, the arms), by bending the joints, and by reducing any unnecessary backswing. As a spectator, observe the techniques used in these skills. Find other uses of the short lever—for example, notice figure skaters' arms when they want to increase the speed of their spin (rotation).

Inexperienced players sometimes have difficulty with reducing the force factor. I have an enlarged knuckle on my right hand courtesy of such a basketball teammate, who literally moved the top two segments of my finger into a new relationship with the rest of my hand (figure 3.4).

Inexperienced players are characterized by excitement and a sense of all-out effort. You should help these players realize that "forcefully" and "quickly" sometimes need to be separated and that skill is involved in knowing when and how to execute each successfully.

Activities

- *Warming up in order to reduce internal resistance.* Here's an activity that can help you understand the importance of warming up: Place your hand in cold water for a minute. Then remove it and try to move your fingers rapidly. Is it difficult at first? Does it become easier after a while? Why?

- *Keeping your joints in a ready position with muscles on a partial stretch.* To test the effectiveness of the ready position, try standing tall with knees straight, and have someone shout a directional command while you attempt to move as rapidly as possible in that direction. What is the first thing that happens to your knees to make this movement possible? Now repeat, but start in a ready position. Do you feel the difference in how quickly you can initiate your response?

- *Involving the stretch reflex.* Avoiding a deep knee bend, first jump and reach as high as you can using a half squat that involves a dip or bounce to induce the stretch reflex. Record your best height. Then, after you have rested, jump and reach again using a half squat but no dip or bounce. Record the best height you can reach. Then compare the two.

- *Using gravity to overcome inertia.* Attempt some movements that you can initiate with a loss of balance. Is inertia overcome rapidly? Can you learn to control this loss of balance so that as much of the fall as possible is used to your benefit?

- *Shortening a lever when maximum force is not needed.* Place a ball on the ground. Ask a player to stand about 10 feet (3 m) away. Moving to the ball, pick it up, and, without using a backswing, throw it to the other player. It might be wise to ask this player to wear a glove.

FIGURE 3.4 The need to complete a task quickly doesn't always call for an increase of force. Sometimes it is more effective to decrease the force in order to complete the task quickly.

• *Shortening a lever to reduce preparation time.* If you are a tennis player or would like a real challenge, try playing the block volley at the net without using a backswing and with the ball closer to you so you can use a short lever. Be sure you know the proper technique for executing the block volley before you begin.

True or False Review Questions

Information about each question is found on the page number provided within the parentheses following the question.

REMEMBER THAT A PARTIALLY FALSE QUESTION IS CONSIDERED FALSE.

Answers to the questions and additional information are found in appendix C.

1. Inertia is a state of stillness that must be overcome to initiate movement (page 23).
2. Overcoming inertia can affect the amount of force that can be generated, but it can also affect how quickly we can accomplish a given task (pages 23, 26).
3. When preparing to initiate movement, you will save time if the joints to be involved are bent and if the muscles to be involved are warmed up and on a partial stretch (pages 23-24, 27).
4. Putting the muscles on a partial stretch evokes the stretch reflex, which can protect the muscles against being overstretched and torn and can help overcome inertia very rapidly (page 25).
5. The use of the stretch reflex to overcome inertia (thus allowing more force to be developed over the available distance) allows a person to complete a task—such as a quick toss to put a player out at first base—more rapidly (page 25).
6. Successful pitchers, field goal kickers, weightlifters, golfers, and tennis players use the stretch reflex when power or force is needed. But a person needs to be aware that evoking the stretch reflex requires additional time during the preparation phase (page 25).
7. An additional way to initiate movement quickly is to use the pull of gravity in an off-balance position. Sprinters, football line players, tumblers, and gymnasts frequently use this source of force to help them overcome inertia and initiate movement quickly (pages 25-26).
8. Shortening levers may get you there faster so that more force can be developed (pages 26-28).

CHAPTER 4

Force

After you initiate an action, you will usually need to continue to develop sufficient force to accomplish the task effectively. The amount of force you will need is directly related to your task objective. Your goal should be to skillfully develop and apply only the necessary amount of force in the most effective way. Involving too much or too little force or applying it inappropriately may actually interfere with meeting your objective. This chapter discusses how to develop and apply force in order to affect movement. But you will need to relate this information to your specific needs in a given situation and make appropriate applications.

Force Development

- *Each contributing body part can add to the amount of force developed (summation)* (figure 4.1).

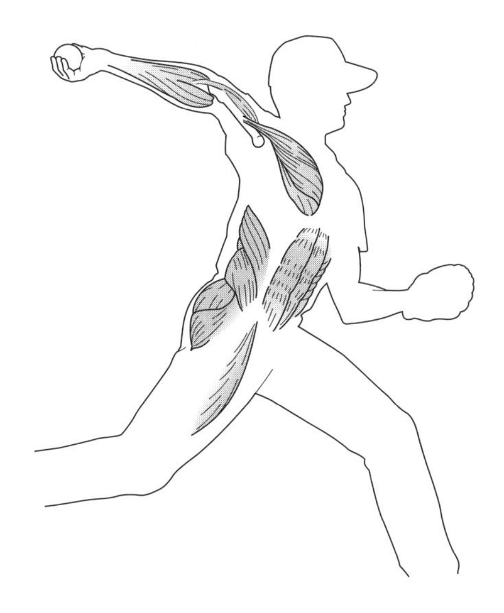

FIGURE 4.1 If maximum force is needed, all potentially contributing muscles must be involved to their maximum extent. For this to occur, the range of motion (ROM) in the contributing muscles becomes important.

To ensure maximum participation of all contributing body parts, you need to use the following that are discussed in this chapter:

1. Opposition (involving opposite sides of the upper and lower body)
2. Full stretching of the involved muscles before using them (this is done through actions such as a full backswing and a full trunk rotation)
3. Stabilization of anchor points
4. Weight transfer
5. Sufficient strength of each contributing body part
6. Full range of motion of all involved joints (available and used)
7. Maximal effort by each contributing body part
8. Follow-through

When less than maximum force is needed, some factors can be appropriately modified:

1. Involved muscles are not put on a full stretch—backswing or trunk rotation may be reduced.
2. Each body part makes less than maximal contribution.

All other factors are usually maintained because of the contributions they make to other aspects of movement, such as balance and accuracy.

If rapid completion of a task is vital to the success of the movement, and if little force is required, then keep the following in mind:

1. Opposition can be considered fairly irrelevant.
2. The backswing should be reduced or eliminated.
3. The range of motion of the involved joints should be reduced (levers shortened).
4. The stretch reflex should not be involved because this would require additional time to invoke.

Other factors are usually maintained because of the contributions they make to other aspects of movement.

- *Each contributing body segment picks up the flow of motion in sequence as the preceding body part has reached its peak contribution.* This produces a continual increase in force development (referred to as sequential involvement), as illustrated in figure 4.2. Force development normally originates from the center of gravity of the body and flows outward toward the end of the involved extremity.

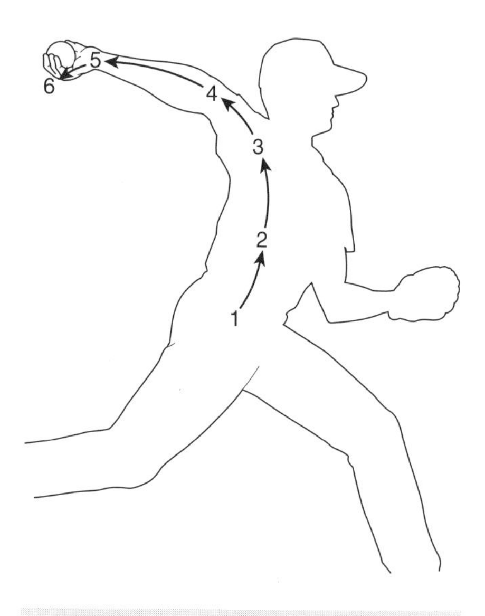

FIGURE 4.2 Force buildup is sequential from the center of gravity to the end of the involved extremity.

To ensure the appropriate sequential buildup of force, these factors need to be present:

1. All body parts in the sequence must be in appropriate condition to participate. Any weak area can interfere with the sequential force buildup and substantially diminish the force that has already been developed.
2. A player must be able to set (stabilize) previously involved parts of the body so that those coming later in the sequence have a firm foundation to pull against in order to make their contribution. (See Skates in chapter 10 for an activity that improves pelvic stabilization.)

Two exceptions occur in the "pacing" of force development. They both involve the need for less than maximal force:

1. The push movement that is involved in a golf putt; a low, short serve in badminton; or a punch volley at the net in tennis is executed with a steady, controlled timing. The outer parts of the extremity do not join in the sequence, so force is not "building."
2. Moves that require absorption of force—such as a bunt, feinting, or a drop shot—require a reversal of the normal buildup of force.

• *For force to develop, some portion of the body must be stabilized.* This stabilized part acts as a brace that moving parts can push or pull against. The stabilized part could also prevent the absorption of force that would occur if there were no stabilization, or it could allow for a gradual or controlled absorption of force when needed.

Many less experienced participants neglect the role that the stabilizers play in force production. Many weekend golfers, for instance, are not aware of the role that the abdominal muscles play in the distance of their drives (figure 4.3). The abdominal muscles stabilize the hips and create the anchor point around which the swing and weight transfer take place. The abdominal muscles play a similar role in soccer, and the shoulder girdle has a stabilizing role in throwing or striking.

Children and other individuals who seem unable to jump, throw, kick, or hit with any degree of success or vigor are frequently lacking in the area of stabilization. Before they can benefit from learning a technique for a specific skill, they must attain adequate development of the involved muscles and stabilizers. This will be discussed further in chapter 5, which covers how to prepare the body to apply the mechanical principles of movement.

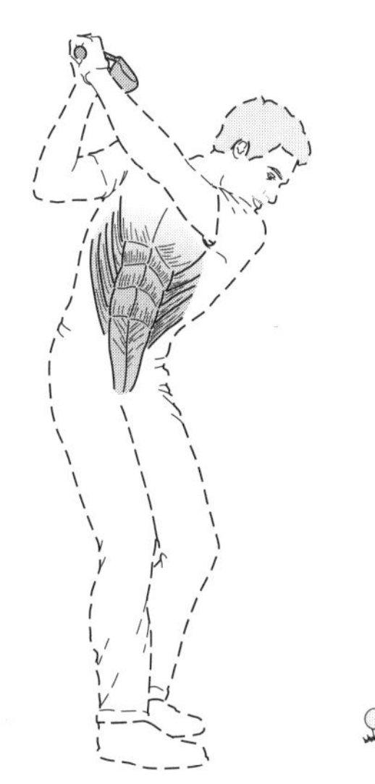

FIGURE 4.3 The abdominal muscles are not only important in force production, but they also stabilize the pelvic girdle and the trunk for greater force in a swing (golf, baseball, soccer, tennis, and so on).

• *Proper weight transfer can contribute substantially to force buildup.*

When you are preparing for weight transfer, your weight may be initially shifted in the direction opposite that of the action phase. This preparatory transfer allows for a more total weight transfer. In addition, this initial shift can pull some of the muscles that will be used during the action phase into a more effective elongated or stretched position, leading to greater force development. This may also stimulate appropriate stretch reflexes, thus helping you overcome inertia rapidly. The shift of weight in the direction you want to apply force during the action phase increases the velocity at which the whole body is moving (figure 4.4).

Many less experienced participants neglect the role that the stabilizers play in force production.

This weight transfer also assists in timing, accuracy, and force development. Transferring the body weight from the back foot to the forward foot allows for a flattening of the swinging arc, permitting you to release the throw, or stroke,

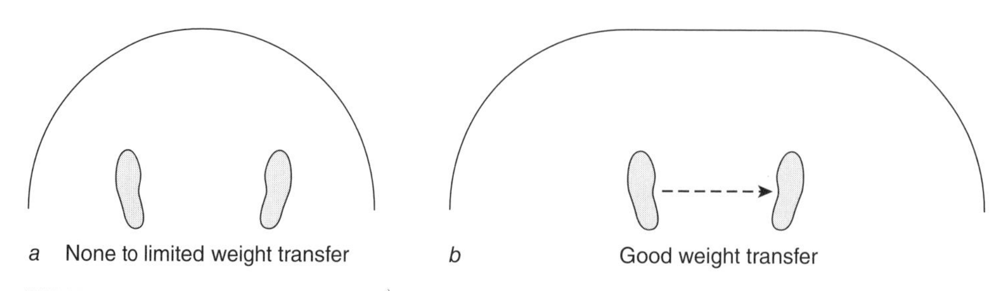

FIGURE 4.4 Can you see the difference in the path of the swing when there is little or no weight transfer and when there is a definite weight transfer?

or strike the object at several points in your forward movement supporting accuracy. Applying force later in the swinging arc can increase the time and distance over which additional force can continue to develop. (See chapter 8 for more information about flattening the swinging arc.)

When there is no weight transfer, the swinging arc tends to be a perfect semicircle; a weight transfer allows you to flatten the swinging arc. This occurs because your swinging arc takes the center of gravity as its center, and as you shift weight, your center of gravity also shifts. So you start making one arc from your initial center of gravity position, and you end swinging around a second center of gravity position. These two arcs overlap, creating a single longer arc with a flattened area caused by the shift of weight and center of gravity (figure 4.5).

Flattening of the swinging arc is extremely helpful to the beginner whose timing may be less than perfect. It also allows the highly skilled player to strike or release later in the flattened arc pattern, increasing the time and distance over which force can be developed before contact or release. The final weight transfer continues over a bent forward knee into the follow-through. This allows for the gradual absorption of force. Follow-through over a bent knee also aids in keeping the center of gravity low and over the base of support, thus avoiding a loss of balance that could negatively affect consistency, accuracy, and preparation for the next move.

- *Force buildup requires time and distance.*

If the muscles to be involved in the action are put on a full stretch before use (as in a preparatory backswing, crouch, or trunk bend or rotation), the time and distance over which force is developed can be increased.

Summation, opposition, trunk rotation, weight transfer, lengthening the lever, and flattening the swinging arc can also increase the time and distance over which force develops. The elimination of any of these, which might occur if the act needs to be rushed, could substantially reduce force development.

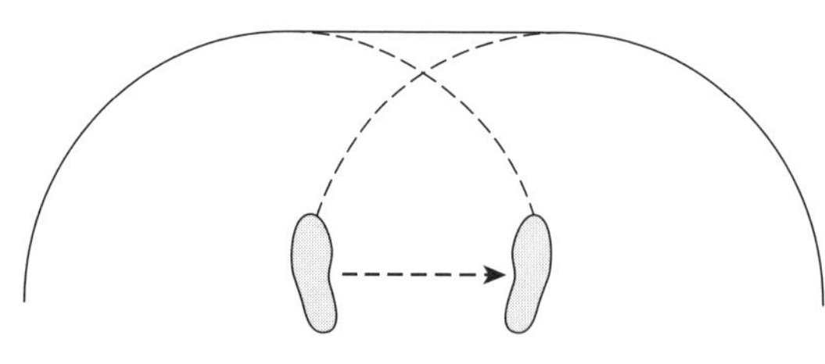

FIGURE 4.5 Can you see how weight transfer allows for a flattening of the swinging arc?

Some participants reduce the time and distance over which force is developed because they believe that rushing contributes to maximum force development.

For these participants, the confusion lies between getting an action done rapidly and getting it done forcefully. This misconception normally fades with experience.

Good range of motion (ROM) can also be an important factor in increasing the time and distance over which a person can develop force. For example, if the ROM in your shoulder joint is limited by tight chest muscles or joint restrictions, you may be limited in your backswing or reach (figure 4.6). Putting the involved muscles on a full stretch before the action phase can increase the time and distance for force development. But this takes time and is not done when a task must be completed quickly but not necessarily forcefully.

Tight quadriceps (the muscles on the front of your upper leg) could reduce the preparatory backswing of your leg, while tight hamstrings (the muscles on the back of your upper leg) could limit how far you can follow through on the kick (figure 4.7).

Good range of motion (ROM) can also be an important factor in increasing the time and distance over which a person can develop force.

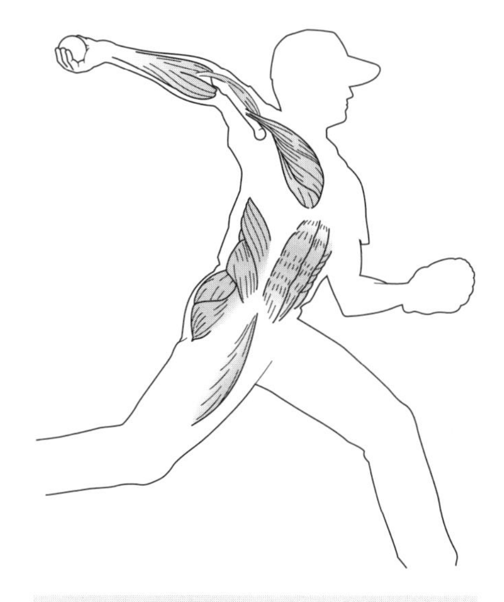

FIGURE 4.6 Muscles need to be long and strong to develop maximum force. A good range of motion (ROM) allows for a full preparatory backswing, while muscle strength is needed in the action phase. Putting the involved muscles on a full stretch prior to the action phase can increase the time and distance for force development. This takes time and is not done when a task must be completed quickly but not necessarily forcefully.

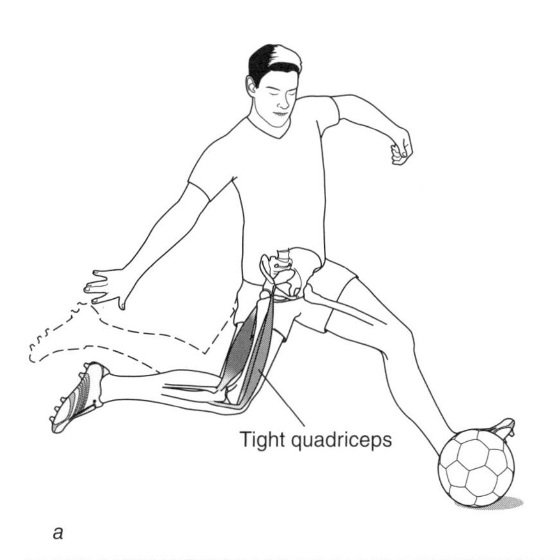

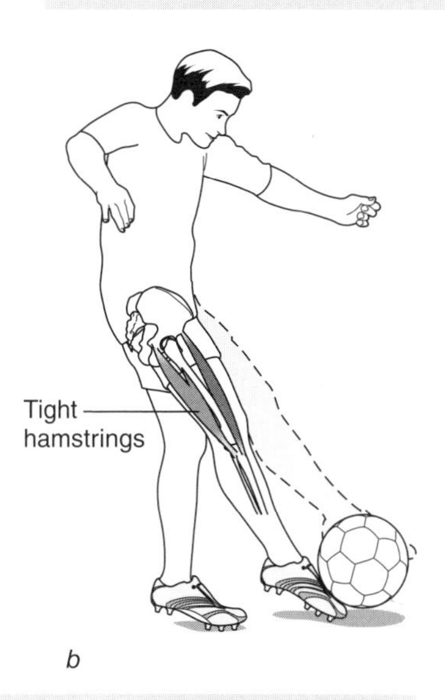

FIGURE 4.7 Tight quadriceps muscles can limit the preparatory backswing and reduce the time and distance over which force can be developed. Tight hamstrings can reduce how far and fast the leg can be kicked forward. Tight muscles can also lead to muscle-tearing injuries.

FIGURE 4.8 Extraneous movement is a very difficult thing to notice. But it tends to reduce force. You may have to begin with the result (lack of force) and work backward to the cause of the extraneous movement.

- *Extraneous movements may reduce or inhibit force production.*

Any movement that does not contribute directly to the movement objective is wasteful and may even require compensating efforts to counteract its effect. One of the goals of practice should be to reduce or eliminate any unnecessary movement. Observe a skillful participant; then watch an inexperienced player. Can you spot the extra movements?

An example of an extra movement can be found among many beginning volleyball players. As the players move the striking arm to hit the ball in the underhand serve, they also move the hand holding the ball forward (figure 4.8). Thus, they literally chase the ball with their striking arm. This makes the contact weak and ineffective. This "overflow" or extra movement may be caused by an inability to use the body parts selectively and separately in a coordinated manner, by a lack of awareness of this extraneous movement, or by a general overall tension level.

- *Noncontributing tension retards the buildup of force and is counterproductive.*

Reciprocal innervation is a process by which dual messages are sent to the muscles—specific muscles are stimulated to contract, while opposing muscle groups receive a message to relax (figure 4.9). This allows the contraction to occur without countering resistances.

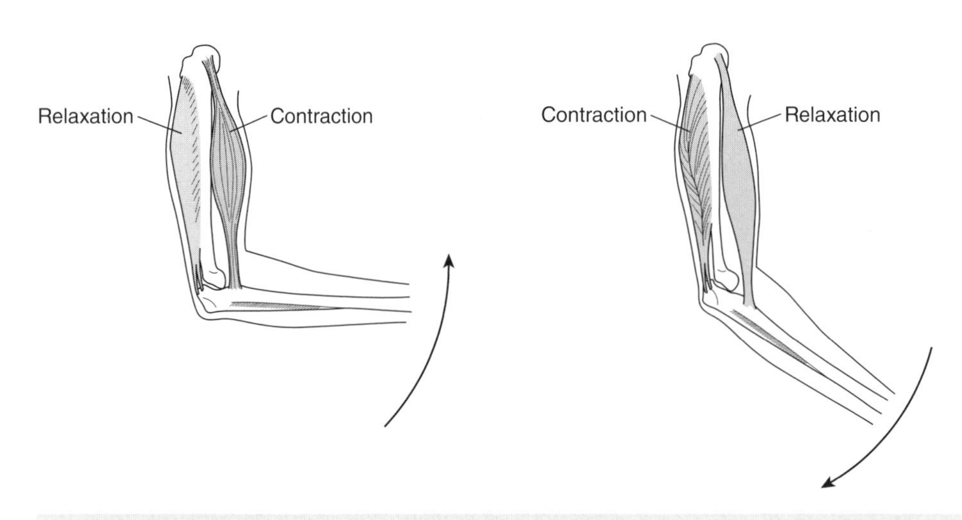

FIGURE 4.9 Muscles work together. As one set of muscles contracts, the opposing muscles should relax (reciprocal innervation) to allow the contracting muscles to work without resistance from the opposing muscle group.

Beginners or participants who have a great desire to succeed often develop a residual, nonproductive tension in the muscles that need to be relaxed. This excess tension inhibits the freedom and flow of movement.

During stressful situations, you can observe even the highly skilled participants attempting to reduce their tension level. Note the swimmer or runner who shakes out before going to the blocks, the golfer who takes a practice swing before the crucial putt, or the basketball player who stretches before a foul shot.

Being aware of the effect of tension can help. You can also learn to relax more. This skill can benefit many aspects of living and should be aided by participation in play experiences.

Force Application

- *Force must be applied in the desired direction.*

To be effective, force must be applied in the desired direction. In striking a ball, for example, the force is applied directly through the center of gravity of the object unless the person is deliberately seeking to apply a spin or rotation.

A very old and important principle of motion involved in the directional application of force is Newton's well-known law: To every action there is an equal and opposite reaction. One way to interpret this statement is to think of the "action" as the application of force and the "equal and opposite reaction" as the resistance or resistive response to this application of force. Another way to state this law might be as follows: To every application of force there is an equal and opposite resistance. Now all this is followed by a result or response, which will depend on the relationship between the amount of force applied and the amount of resistance met.

Here are two possibilities: (1) The object of resistance is movable, and the force is sufficient to move it. The result is that the resisting object moves in the direction of the application of force. Examples might include hitting a ball, pushing a car, and throwing or kicking. (2) The object of resistance is immovable (the earth or a wall), so the force applied cannot move it. The result is that the source of force moves in the direction opposite to the application of force. Examples might include jumping, running, walking, or hitting a ball against a wall. If the resisting object and the source of force are both movable, they will both respond to the force and move away from each other. Examples occur in billiards or with a bowling ball and pins.

Thus, if you want to move *forward,* you need to push *backward* in swimming or running. If you want to jump *upward,* you must push *down.*

Can you see how starting blocks help direct the force in a more desirable direction (figure 4.10)?

Furthermore, if you want an object to go *up,* you must hit *under* it (volleyball set); if you want it to go *straight ahead,* you must hit *behind* it (line drive); if you want it to go *down,* you must hit on *top* of it (tennis serve, volleyball spike). We can also use this information to analyze errors. An object that does not go upward enough was not hit underneath enough, and one that goes upward too much was hit too far underneath. To correct these errors, emphasis must be put on the appropriate contact points. (See figure 4.11; also see the section on contact points in chapter 8.)

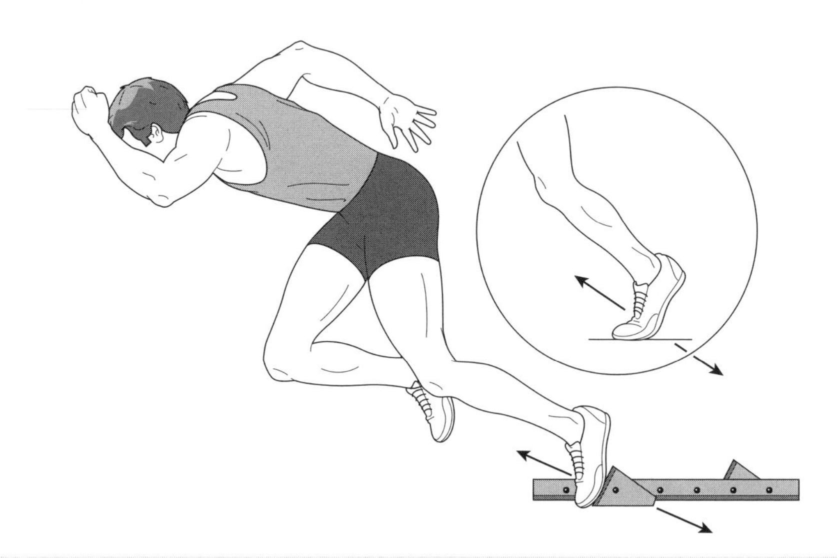

FIGURE 4.10 When a person uses starting blocks, Newton's law applies: "For every action there is an equal and opposite reaction."

FIGURE 4.11 Which of these contacts will take the ball over the net?

Application of force in any direction other than the desired one can lead to an undesirable result. A tennis stroke can be very forceful, but if the ball is hit upward or too far to the side, this force can lead to failure (out-of-bounds play). In swimming, where the force is applied throughout the action phase, force application that counters the desired direction of propulsion is not only wasteful but may require additional compensatory effort or may lead to weaving or bobbing.

- *When the goal is maximum distance, the angle of release, impact, or takeoff may determine the effective application of the developed force.*

Because many projectiles have a predetermined parabolic path, it has been possible to study the effect of various angles of release, takeoff, and impact. A 45-degree angle will normally provide maximum distance when the initial and touchdown points are on the same level, as in a golf stroke or placekick (figure 4.12).

If a projectile, such as a ball, or the center of gravity of a human projectile travels from a point higher than it will land—as in a home run, punt, shot put, javelin throw, or broad jump—the angle should be reduced slightly to compensate. Of course, if you had to golf from a lower level to a higher elevation, the reverse would hold true. If wind conditions or other airflows affected the object, this principle would also have to be modified according to these variables.

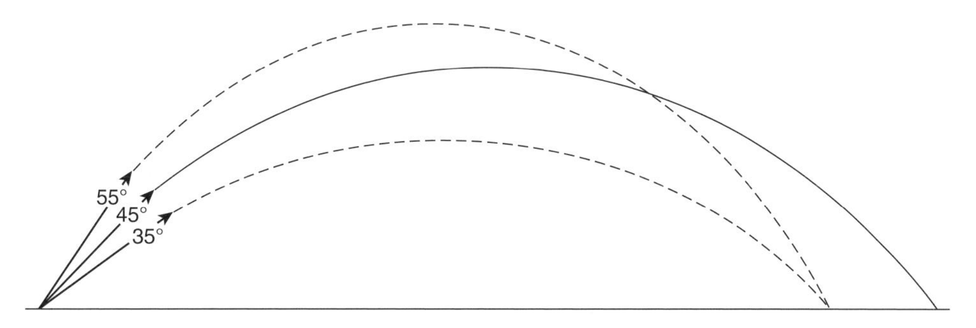

FIGURE 4.12 Various angles of release, takeoff, or impact will create different paths of travel and landing.

• *A good follow-through ensures maximum velocity and force at the point of release or impact.*

Many people wonder how the follow-through, which follows the application of force, can have an effect on force application. A complete and extended follow-through ensures that the slowing down process will not be initiated too early during the final part of the action phase. This means that maximum velocity is still available at impact, release, or other time of need. The effects of the very subtle error of slowing down are perhaps more obvious if you observe inexperienced runners who, seeing the finish line, slow down as they approach it.

A complete follow-through also allows for a gradual absorption of force, preventing injuries (such as muscle tears) or a loss of balance. Thus, a good follow-through plays several important roles in the effective application of force.

Absorbing Force

Although there are times when we need to develop force, there are also times when we need to absorb a force. A person may need to absorb a force in order to avoid injury, to stop a movement, or to maintain balance or control. When these situations occur, knowing how to absorb a force may mean the difference between success and failure or injury.

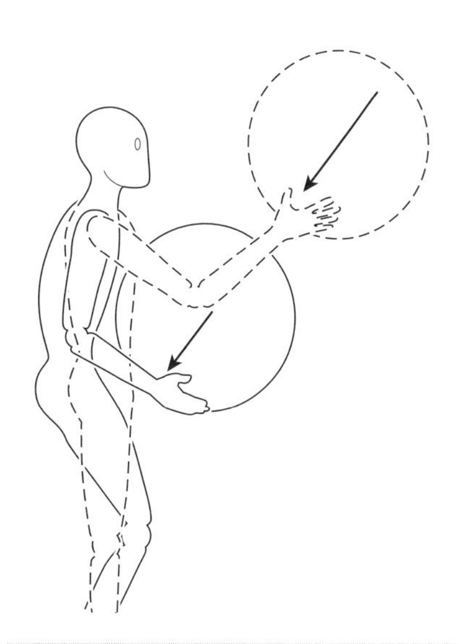

FIGURE 4.13 We can reduce force by giving with the joints.

Force can be reduced by *giving with* or *spreading* the force. *Giving* with the force slows it down gradually over time and distance. Our joints allow us to accomplish this by bending (giving gradually) as we receive the force (figure 4.13).

Spreading the force out over a larger receiving area means that no single part of the area takes as much of the total force. The larger the area receiving the force, the less the impact force will be on any part of that area. For example, a catcher's mitt acts like a cushion and as it compresses, it continues to increase the size of the area of impact (figure 4.14).

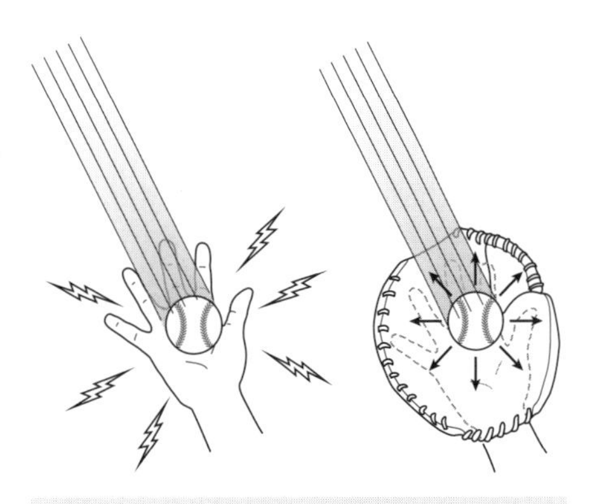

FIGURE 4.14 Some sport equipment helps SPREAD the impact force.

Note the difference in size of the impact area in figures 4.15 and 4.16. Which would tend to be less injurious?

In landing, you would need to give with your ankles, knees, hips, and vertebral column by allowing them to bend in order to control and absorb the force of your weight. However, if you fell while landing, you would want to have the larger and fleshier parts of your body take the impact. Each of these factors will increase the spreading of the impact force.

Can you see how a ball that is soft would increase the impact area as it compressed on contact? Would this hurt less in a game such as dodgeball compared to a ball that was harder? Can you also see why this ball would be less apt to roll or bounce away? Or how a good pair of tennis shoes could help spread force in stopping? Why would pushing your feet downward against the floor and letting your knees bend help you stop more effectively? How does leaning backward give you more time and distance over which to absorb force? Can you see how friction could help spread the force? Have you seen judo or karate participants roll to spread the impact of landing over a large body area? Think about how different parts of the shoulder pads in football help cushion and spread the force of impact.

Force can be reduced by *giving with* or *spreading* the force.

When sliding into a base in softball or baseball, a player wants to first avoid being tagged, but the player also needs to absorb force in order to avoid sliding beyond the base. Therefore, when sliding into a base, the player needs to have as much body surface as possible come in contact with the ground. This spreads the force of impact and also slows the player down.

Can you see how a stride stance (forward and backward) could at times help you to absorb a force that you are receiving? In sports such as bowling, could a

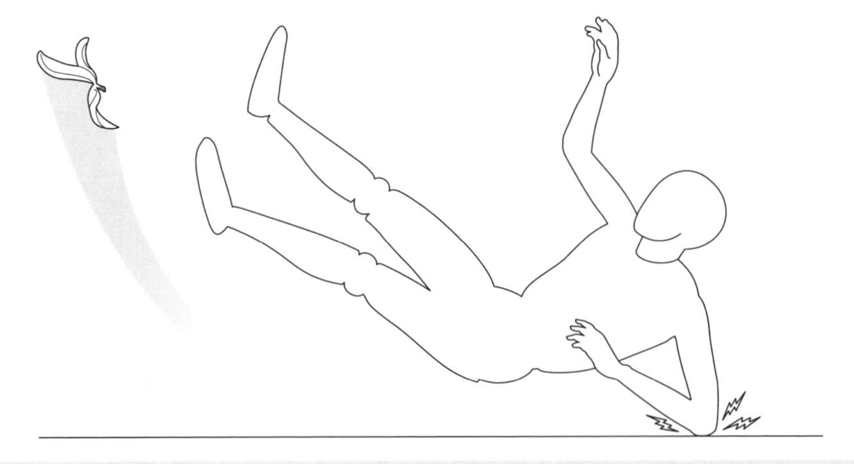

FIGURE 4.15 Would this landing spread the force of the impact and be apt to prevent injuries?

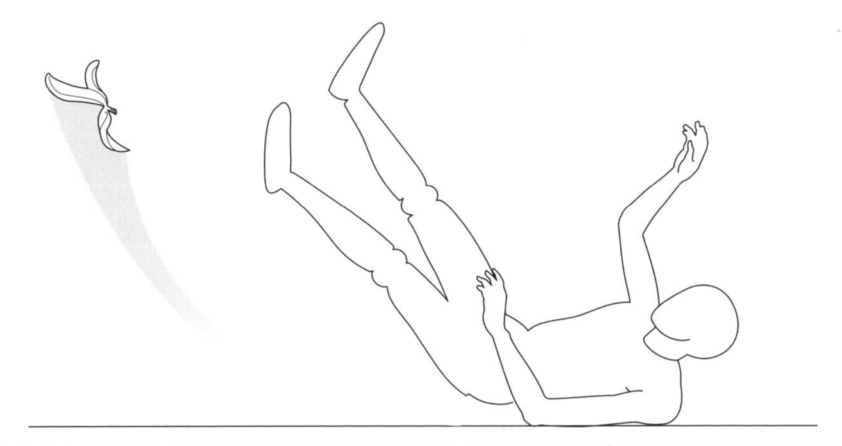

FIGURE 4.16 In this figure, can you see that the larger, softer part of the body, which is receiving the impact, both spreads the impact force farther and will continue to increase the size of the impact surface as it acts as a cushion and compresses more than the landing shown in figure 4.15?

forward and backward slide or a stride position over a bent forward knee allow the force to be gradually absorbed on the follow-through and allow you to maintain your balance? Refer to the game matrix in chapter 10 (page 100) for several activities that will help players learn to absorb force.

If you were standing in a bus that made several starts and stops, how would you make it possible to give with and spread the force that you receive? How do wrestlers absorb force? How does a hockey player control a fast-moving pass? How does a baseball player control the force of a pitch when bunting?

Try to think of a variety of situations in which force absorption allows for further control or the avoidance of injury. Can you see how a good range of motion could help in the absorption of force over time and distance and could possibly prevent small tears within the muscles? Would the strength required to slow the force down gradually have an important role?

Can you see why runners and joggers are discouraged from running on very hard surfaces with little give? Do basketball floors give at all?

Opposition

Opposition (the involvement of the opposite sides of the upper and lower body) plays several roles in relation to force (figure 4.17). Opposition makes it possible to have the following:

1. A longer preparatory backswing as well as a more complete follow-through, which increases the time and distance over which force can be generated and applied
2. More total muscle involvement through summation and sequence
3. A full stretch of the trunk muscles before their contraction
4. Greater transfer of weight, allowing the body weight to participate fully in force development

FIGURE 4.17 The use of opposition in throwing makes a greater backswing possible and allows for more muscle involvement when greater force is needed.

5. A better balanced position throughout the movement, creating a more stable base from which the mover can push to increase force
6. Greater ease in adjusting the center of gravity over the base of support for good balance
7. A position that allows the person to absorb a received force and also to be prepared for the next move (For further practice in absorbing force, see Super Sox in chapter 10 [page 119].)

Failing to use opposition is a common error among the very young, the inexperienced, or those who have developed their own style. It is so common among the young and inexperienced that it is considered part of the growth and development process. If players have not developed the use of opposition naturally through experience, they should be encouraged to use this position. At first the change may feel awkward, but an individual can adjust with patience and perseverance. After players have begun to feel comfortable with this new position, have them attempt several throws for distance. Can they observe any measurable difference?

Although the use of opposition is important, there are some exceptions in specific sports or for specific reasons, such as a quick toss to the base to catch the runner. Some badminton players have preferred not to use opposition in the low, short serve in order to protect more easily against a quick backhand return. Fencers do not use opposition because they prefer to make minimal contact surface available to their opponent. Dart throwers do not use opposition in order to avoid body rotation, which could reduce accuracy.

Lever Length

The longer the lever, the greater the force potential at the end of it. Lengthening the lever, of course, may be of no value if the length prohibits physical control of the lever, reduces the participant's ability to move it through a complete pattern, or reduces the participant's ability to complete the act in the time available.

The length of the lever affects both rotary speed and force. A shortened lever can be brought around or rotated more rapidly but will normally have relatively less force; the lengthened lever may be slower and more difficult to rotate, but it will be able to build up more velocity over the additional distance. Thus, a full reach (not crowding the ball or not choking up on the striking instrument) may help a player hit a home run or have a more forceful tennis stroke, but it will require a longer time to execute. This creates no time problem in movements such as the tennis serve or a golf drive, but a fast pitch might get by you during the time required to take the "long" swing. When this occurs, you may choose a shorter lever with less force so that you can get around in time to meet the ball. If the oncoming object is travel-

The longer the lever, the greater the force potential at the end of it.

ing fast, you may be able to use the momentum of the oncoming ball to contribute to your force rather than develop force through the use of a long lever.

A shortened lever can be brought around or rotated more rapidly but will normally have relatively less force.

If your objective is to react quickly or increase the speed of rotation—such as in a net play in tennis, a concealed bunt, sprinting, a multiple somersault dive, a skater's spin, a short throw, or a quick kick—using a shorter lever would be more effective. But if your objective is force and if you have sufficient time, distance, and strength to accomplish the specific task, you could be more successful if you applied a longer lever. This would be the case in a baseline tennis stroke, a home run, a "ring-a-bell" game at a carnival, a long throw, a long kick, or a golf drive in which distance is an important factor.

In relation to lever length, can you answer the following questions? If you wanted to do as many somersaults as possible in a dive, how could you most effectively accomplish this? If you wanted to make a dive appear slow and graceful in the air, how could this be accomplished? Can you see any application of this to gymnastics?

What would happen if skaters drew their arms into their body, shortening these levers when they were spinning? Does this apply to a pivot in basketball?

In tennis, net play requires rapid reactions but less stroke force. How would your net stroke differ from your baseline strokes? Note that because the force needs are reduced, the length of the backswing can also be reduced.

In making the choice of whether to use a longer or shorter lever, a player needs to realize that a quick movement may consume proportionately more energy for the amount of movement accomplished. This is because a quick movement does not take full advantage of momentum and the increased time and distance available to develop force. If people attempted to sprint a marathon, they would probably find themselves exhausted early in the race. Each situation requires the participant's consideration of the factors involved—this is part of the fun of participation.

Observations

Observing unskilled individuals can be helpful. While watching these participants, see if you can find any common faults in their force development. When they are trying for maximum force, do they use opposition? Do they put the muscles to be used on a full stretch? Do they shorten their backswing and rush the preparatory phase? Do they use any unnecessary movements? Does their motion flow, or do they hesitate at any time during a movement, thus breaking the sequential timing of the involved body parts? Can you see a real and complete weight transfer? Do they use a complete follow-through? Can they adapt to varying task needs (objectives)? Can they develop only the amount of force that is needed? Can they reduce the time needed to complete a time-limited task such as a short throw or a bunt? Can they reduce or eliminate the backswing if it is not needed? Can they effectively shorten as well as lengthen a lever for an appropriate result? Are they tense? Can they set (stabilize) various parts of their body to help them move well? Do they apply force in the desired direction?

Observe several joggers. Try to find one who toes out severely when running. Can you see the body rotation from the force being pushed out to the side rather than backward? Can you relate this to Newton's law of action and reaction? Try to find a jogger who has a great deal of up and down motion. Can you see that this person is pushing downward more than backward? If these joggers are out for the exercise, they are meeting their objective. But if they want to use their efforts most efficiently, do their best, or avoid stress on body parts, they may need to change their style.

If you are able to observe a shot-putter or a golfer going for distance, you should attempt to calculate the angle of object projection. Remember, straight up is 90 degrees, and an angle that is one-half of this would be 45 degrees.

Activities

- *Summation:* Play catch. In attempting to throw for distance, use various amounts of backswing. Compare your results.
- *Length of lever:* Find a piece of Styrofoam large enough to allow nails to enter it to approximately 2 or 3 inches (5.1 or 7.6 cm); the piece of Styrofoam should be about 18 inches (45.7 cm) long and at least a couple inches wide. Attach a weight to a piece of string roughly 16 inches (40.6 cm) long. Attach this to the top of your piece of Styrofoam. Place two nails into the face of the Styrofoam, one approximately 15 inches (38.1 cm) from the top and one roughly 8 inches (20.3 cm) from the top (figure 4.18).

Now, holding the string parallel to the ground, let it swing free and hit the nail at 15 inches. (*Note:* If you are having difficulty with the weight missing the nail, place the whole device next to a wall and let the string swing along the wall.) Shorten the string to 8 inches and let the weight swing. Is the nail hit by the weight at 15 inches driven deeper than the nail hit by the weight at 8 inches? What does this tell us about the force developed at the end of a longer and shorter lever?

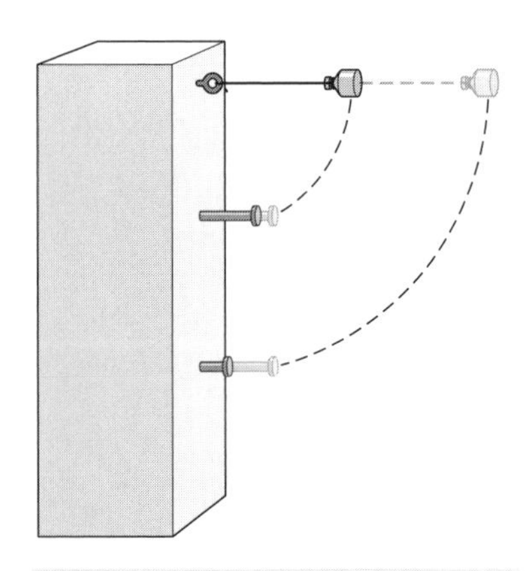

FIGURE 4.18 From this experiment, can you determine anything about the force developed by a longer versus a shorter lever?

Put an eyelet screw or nail into a wall. Pass the end of the string through the eyelet or over the nail. Now start the weight swinging free. As it swings, pull up on the string (shortening it) and then let it again swing long. What happens to the speed at which the string is swinging? Does this say anything about the speed of rotation of a long and short lever? Can you think of some situations in which each might apply?

- *Weight transfer:* Set up a batting tee. (See appendix B, pages 128-129 for information on how to make your own batting tees.) Hang a sturdy backdrop on the clothesline so you can hit into it. Bat the ball several times using a complete weight transfer each time. Listen to the impact, feel the force. Now, trying to keep your weight centered between your feet (no weight transfer), take several more hits. Is there a different feeling of force? Does the impact sound different? If you had more difficulty hitting the ball or hitting it straight ahead when you did not transfer your weight, can you relate this problem to the pattern of the swinging arc in the two different styles? In which style does the swinging arc tend to flatten more? Why? You might like to try the Batting (or Golf) Tee Challenge in chapter 10. This activity supports weight transfer, flattening the swinging arc, and a good follow-through.

- *Opposition:* Play catch using opposition and not using opposition. Compare the distance of your throws.
- *Angle of projection:* Try this experiment outdoors when there is no wind. Take a garden hose and turn on the water. Holding the nozzle at ground level, aim the stream of water at various angles. At which angle does the water go the farthest? If you can't tell, set up some cans as targets. Hold the nozzle about waist high. Can you perceive any change in the angle of projection needed to get the greatest distance at these two different heights? To avoid waste, you might carry out this experiment on a lawn that needs watering.
- *Extraneous movement:* Try a task that you are fairly proficient at, but use your nondominant hand or foot. Do you notice any extra movement or tension in other parts of your body? Can you become more relaxed and proficient with practice? Try a skill, such as tying your shoelace, with one hand. Do you feel tension or extra movement occurring anywhere in your body? Ask a friend to do this, and observe him.
- *A thought problem:* You are running and want to change directions. You bend your knees to absorb the force and maintain your balance. In using gravity to help you "fall" in the new direction, where would you shift your center of gravity? Which way would you be leaning? Now which way would you push against the ground to go in your new direction? Try it. Does it work? You might consider using the game of Freeze in chapter 10 for this experiment.
- *A thought question:* In sports such as bowling and fencing, could a forward and backward slide or a stride position over a bent forward knee allow the force to be absorbed on the follow-through and allow you to maintain your balance?

True or False Review Questions

Information about each question is found on the page number provided within the parentheses following the question.

REMEMBER THAT A PARTIALLY FALSE QUESTION IS CONSIDERED FALSE.

Answers to the questions and additional information are found in appendix C.

1. Opposition refers to reciprocal innervation (pages 29, 34, 39-40).
2. If rapid completion of a task is vital to the success of the movement, and if little force is required, then the stretch reflex should not be involved (page 30).
3. Force development normally originates from the center of gravity of the body and flows outward toward the end of the involved extremity (page 30).
4. All involved body parts must be in the appropriate condition to participate. Any weak area can interfere with the sequential buildup of force and can substantially diminish the force that has already been developed (page 30).

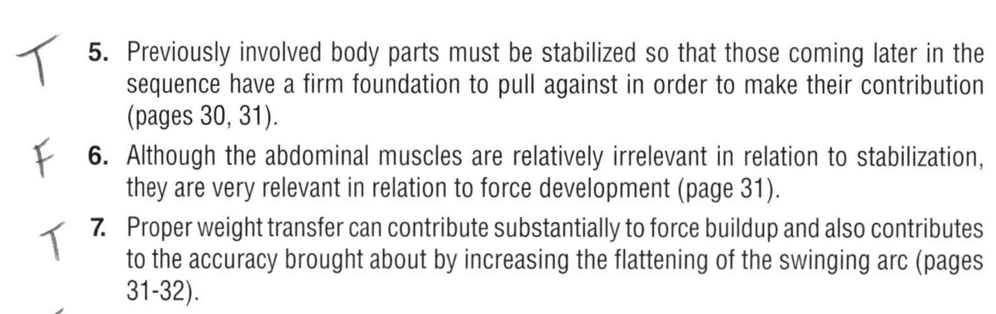

5. Previously involved body parts must be stabilized so that those coming later in the sequence have a firm foundation to pull against in order to make their contribution (pages 30, 31).
6. Although the abdominal muscles are relatively irrelevant in relation to stabilization, they are very relevant in relation to force development (page 31).

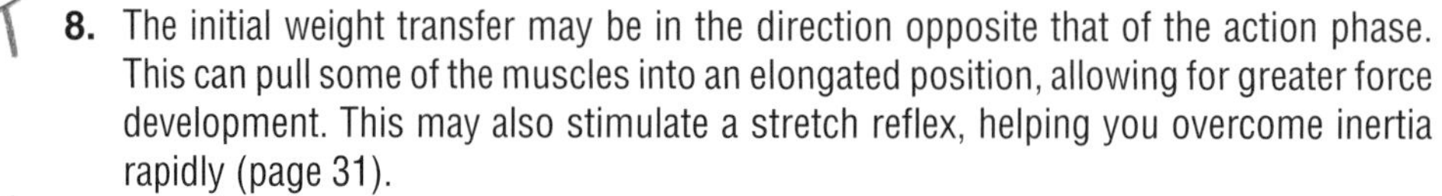

7. Proper weight transfer can contribute substantially to force buildup and also contributes to the accuracy brought about by increasing the flattening of the swinging arc (pages 31-32).
8. The initial weight transfer may be in the direction opposite that of the action phase. This can pull some of the muscles into an elongated position, allowing for greater force development. This may also stimulate a stretch reflex, helping you overcome inertia rapidly (page 31).
9. Flattening the swinging arc benefits the advanced player, but not the novice (page 32).
10. The final weight transfer continues over a bent forward knee. This allows for the gradual absorption of force. It also keeps the center of gravity low and over the base of support, helping the person avoid a loss of balance that could negatively affect consistency, accuracy, and possible preparation for the next move (page 32).
11. A full stretch on the preparatory backswing, the inclusion of the stretch reflex, summation, sequential "flow," opposition, trunk rotation, weight transfer, shortening the lever, a good range of motion (ROM), flattening the swinging arc, and a good follow-through each increases force development (pages 32, 33, 40-41).
12. A good follow-through ensures maximum force and the gradual absorption of force (pages 32, 37).
13. There are times when a person needs to absorb a force in order to avoid injury, stop a movement, or maintain balance or control. Force can be reduced by *giving with* or *spreading* the force. Giving with and spreading the force are both done by bending the joints as the impact is received (pages 37-39; also note figures 4.13 and 4.14).
14. When an object or the body compresses upon impact, the area of impact is increased. This allows the force to be absorbed over a larger area (pages 37, 38).
15. When force is not a factor, opposition might not be used, such as in fencing, darts, or a low, short serve in badminton (page 40).
16. Lever length affects both rotary speed and force. A shortened lever can be brought around or rotated more rapidly for greater force. The lengthened lever may be slower and more difficult to rotate, but it will be able to build up more velocity over the additional distance (pages 40-41).
17. A short lever can be brought around more quickly, but the longer lever will produce more force (pages 40-41).

CHAPTER 5

Preparing Your Body to Apply Mechanical Principles

Knowledge of the mechanical principles of movement makes us more able to use our bodies effectively, but we also need to prepare our bodies to apply these principles more effectively. In the process of this preparation, we may reap the benefits of feeling better physically, increasing the physical potential of our body structures, reducing the possibility of injuries, enhancing our responsiveness to the environment, avoiding unnecessary joint trauma, and improving or maintaining good alignment.

Although sports and play can contribute very positively to the development of the body, researchers have found that our physical needs may not be totally met through specific participation and that we may need to supplement our activity program with some additional physical preparations. These may include strengthening the stabilizers of the pelvic and shoulder girdles, developing and maintaining muscle tone in the antigravity or postural muscles, and stretching muscles that tend to tighten (shorten).

Improving Posture

Poor posture or alignment is frequently a good indicator of your physical needs. Are your shoulders forward? This could be caused by tight chest muscles or weak upper back muscles or both. The shoulder joint is the pivotal point around which arm movements occur. Good range of motion of this joint is necessary to allow the time and distance needed to build force. And sufficient strength is required to stabilize this part of the body, which is the foundation that is pushed or pulled against for all arm movements.

Although sports and play can contribute very positively to the development of the body, researchers have found that our physical needs may not be totally met through specific participation and that we may need to supplement our activity program with some additional physical preparations.

Does your back sway? This could be caused by tight iliopsoas muscles or weak abdominal muscles. The pelvic girdle is another important stabilizing area. Good abdominal development is necessary to hold the bony pelvic region in position. This stabilization is necessary for four very important mechanical functions:

1. Balance can be better controlled if the area where the center of gravity is most frequently found can be held firmly and if adjustments can be made quickly to any subtle or sudden changes.
2. The rotation of the trunk is in part carried out by the abdominal muscles. These muscles also help to hold the pelvis, allowing these trunk movements to occur.
3. All forceful moves of the legs are dependent on having a stable pelvic foundation to push and pull against.
4. The forceful movements of the upper limbs normally originate in the pelvic region of the trunk, requiring a double stabilization—first that of the pelvic area and then that of the shoulder girdle.

Are your knees hyperextended? This could be the result of a swayback or tight muscles in the back of the upper legs (the hamstrings).

Do you slouch? Are you "hanging on your ligaments"? This could be caused by a lack of muscle tone in your antigravity or postural muscles.

Good posture or alignment is thought of as a position or positions of the body, but to achieve the various "good" postures, a person must attain a set of important body conditions that lead to a balance of strength and length of opposing muscle groups. If these conditions exist, a person tends to have good alignment. But even more important, the person has greater freedom and control to move well. In addition, the person is ready to spring into action.

Strengthening and Toning Muscles

Gravity continually pulls us toward the earth. This is evident when we are off balance and fall. To counter this pull, we need muscular strength and endurance in the form of muscle tone. To prevent a fall, we must contract some of our muscles to maintain our balance and must then realign our center of gravity, placing ourselves in a balanced position. This is one reason why it is so important for adults to maintain strength and endurance, why children should develop strength and endurance, and why people who are prone to falls should enter into a developmental fitness program.

An indication of a well-conditioned muscle is *muscle tonus.* When a muscle has good tonus, a number of fibers of the muscle are continuously in a state of contraction. The many fibers within the muscle share this task as a team, rotating which fibers are in contraction at any one time. This prevents fatigue and gives the muscle a feeling of firmness, while maintaining a state of continuous readiness. The "ready" condition allows an immediate adjustment to subtle or sudden changes and increases the person's body control. This condition of muscle is maintained through continual healthy involvement.

Even standing and sitting require us to constantly counter the pull of gravity. If we have poor muscle tone, we will tend to droop, slouch, and "hang on our ligaments." This further increases fatigue, puts stress on joint structures, and reduces our ability to make effective and rapid movement responses.

Because fitness and good alignment require maintenance of the total body, some areas or muscles of the body may need specific attention to make sure they remain effective, contributing members of the muscular system. Two areas that play a vital role in stabilization and may require additional attention to maintain the needed strength, endurance, and muscle tone are the upper back and shoulders and the abdominal muscles (figure 5.1).

Observe people who droop because of poor muscle tone. Can you see why they would have more difficulty in adjusting their balance? Why would it take them longer to respond to an initial loss of balance, and why would they be more apt to lose their balance and fall? Why might these people be more accident prone and less physically skillful? Why are they less ready to move? Can you see that they have little "give" left to stop certain moves or receive force from certain directions?

Can you also see how a lack of muscle tone in the upper back and shoulder muscles could result in forward, drooping shoulders? Consider how weakness in the upper back and shoulder muscles could affect the preparatory backswing in a throw or a racket stroke. Think about how a lack of strength in the shoulder girdle could reduce a person's ability to stabilize that area so that it can act as a brace to be pushed or pulled against during a movement of the upper limb, as in hitting and throwing.

Can you see how a lack of muscle tone in the abdominal muscles could result in a swayback? A lack of strength and endurance in the abdominal muscles could also reduce a person's ability to stabilize the pelvic area that supports force development and accuracy. Stabilizing the pelvis allows it to act as a brace that can be pushed or pulled against during a movement of the lower and upper limbs.

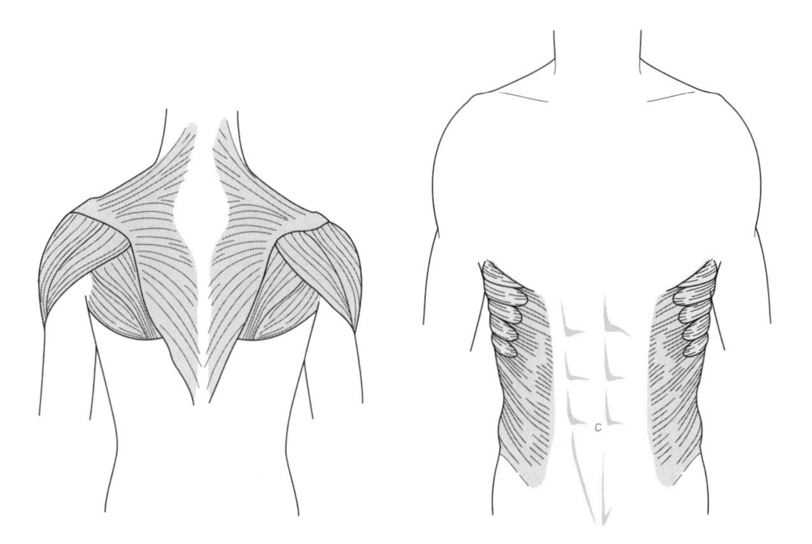

FIGURE 5.1 How would a lack of muscle tone in these muscles affect the posture and performance?

These muscles also support the adjustment of the center of gravity over the base of support to maintain balance. Keep in mind that adjusting the center of gravity plays a vital role in balance and that balance is the foundation of all movement.

FIGURE 5.2 Could a basketball player reach higher to recover rebounds if tight chest muscles were periodically stretched to keep them lengthened?

Correcting Muscle Tightness

While active participants attempt to maintain adequate muscle strength and tone, they should also recognize the need to include stretching in their fitness program to avoid muscle tightness and to ensure that an adequate range of motion (ROM) is available in the various joints of the body.

Our movement takes place around joints through a system of opposing muscles. Muscles must be strong to create and stop motion (absorb force), but they must also be sufficiently long and elastic to allow for freedom of movement and a good initial position from which to move. Without periodic stretching, this freedom and range of motion can be gradually diminished (see figures 5.2 and 5.3).

Muscles that are heavily used tend to form connective tissue within them. Part of this is actually a form of scar tissue that develops as a result of the many microscopic tears or "small accidents" that normally occur during active involvement. If you have had the opportunity to compare the chewiness of the meat (muscle) of wild game and that of an animal that has been allowed only limited activity, you probably noticed the difference in the "toughness." The latter normally needs less tenderizer.

Because connective tissue shortens if not periodically stretched, a heavily used muscle may actually shorten, causing tightness and a loss in range of motion. This condition can also limit the time and distance over which force can be developed.

Can you see how this could reduce force potential, reduce the "give" available in receiving a force, and increase the possibility of injuries that could lead to further tightness and joint restrictions? This condition of tightness can pull the body out of alignment and put increased stress on the involved joints.

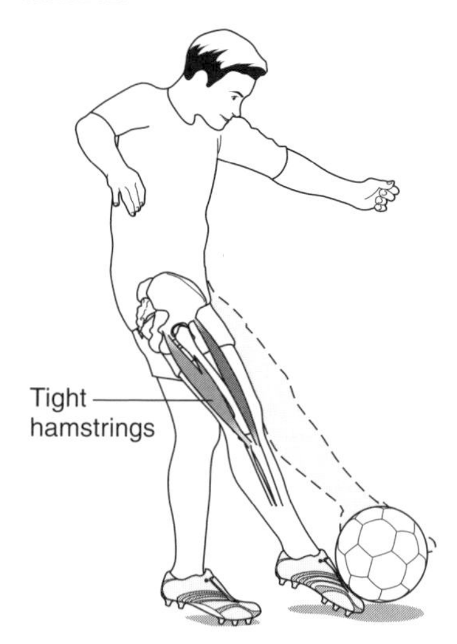

FIGURE 5.3 Could keeping the hamstrings lengthened make it possible for a kicker to kick farther? Why?

Athletic coaches realize the value of stretching and have increased the emphasis on stretching in their conditioning programs. This component has also been increased in many fitness programs.

Muscles can be lengthened—or their length can be maintained—by periodic, *slow* stretching. A muscle that is shortened (or could become shortened) should be brought to its full length and held on the stretch for several seconds. This should be done several times a week.

Note that ballistic or forceful bounce stretching is no longer used because of its injury potential (minor muscle tears) and the involvement of the stretch reflex, both of which can reduce the desired effect of lengthening the stretched muscle.

Figure 5.4 shows the four most common locations of muscle tightness. These are the pectoralis major (upper chest and front part of the shoulder area), the iliopsoas, and the large muscles on the upper front (quadriceps) and the upper back (hamstrings) of the thighs.

While active participants attempt to maintain adequate muscle strength and tone, they should also recognize the need to include stretching in their fitness program to avoid muscle tightness and to ensure that an adequate range of motion (ROM) is available in the various joints of the body.

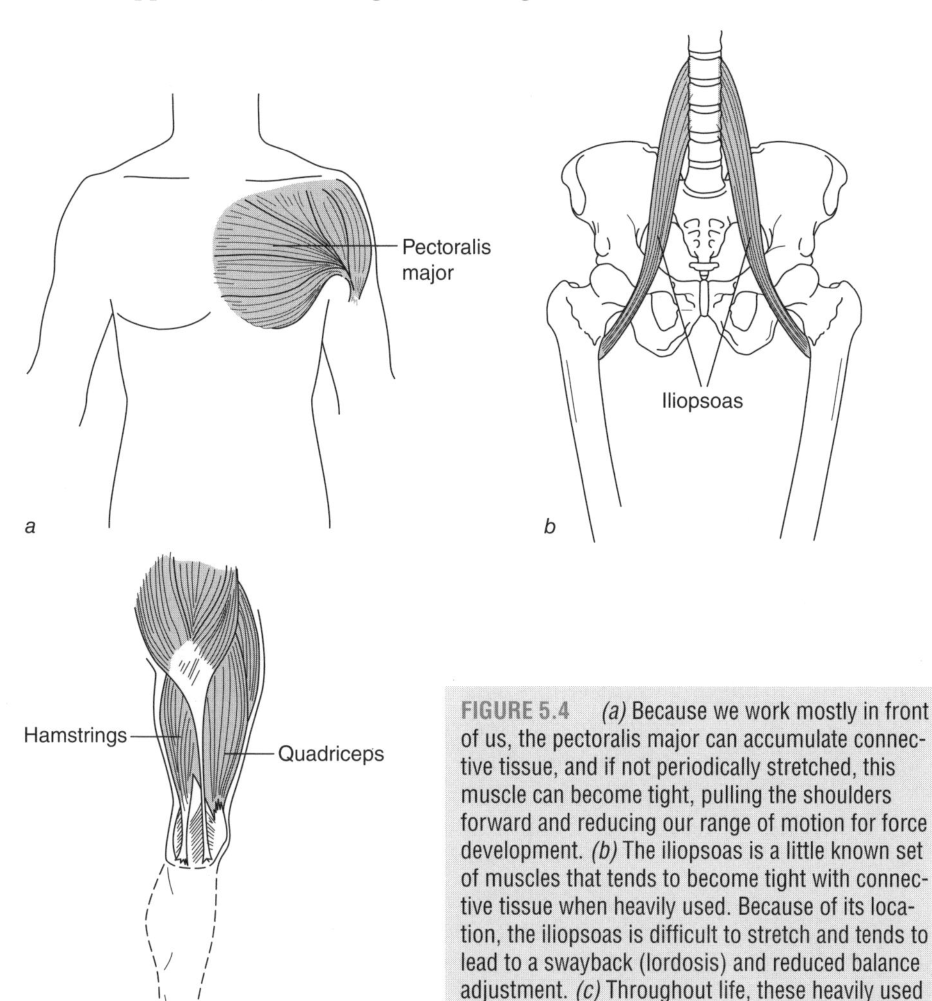

FIGURE 5.4 *(a)* Because we work mostly in front of us, the pectoralis major can accumulate connective tissue, and if not periodically stretched, this muscle can become tight, pulling the shoulders forward and reducing our range of motion for force development. *(b)* The iliopsoas is a little known set of muscles that tends to become tight with connective tissue when heavily used. Because of its location, the iliopsoas is difficult to stretch and tends to lead to a swayback (lordosis) and reduced balance adjustment. *(c)* Throughout life, these heavily used muscles develop connective tissue that restricts their range of motion and thus restricts actions that involve these muscles.

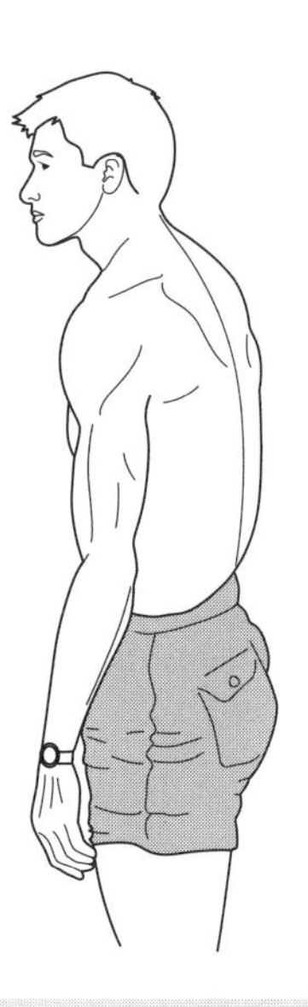

FIGURE 5.5 Which muscles need to be stretched and which muscles need to be strengthened to improve this individual's posture and provide full use of the shoulder joint?

The tight chest muscles (pectoralis major) pull against the upper back muscles, which may be weak, causing forward shoulders (figure 5.5).

A tight iliopsoas pulls the vertebral column forward, and if the abdominal muscles are not sufficiently strong to counter this pull, lordosis (swayback) can result (figure 5.6).

The weak muscles must be strengthened, and their corresponding tight muscles must be lengthened. Again, posture (alignment) provides a good way to evaluate the balance between the length and strength of opposing muscle groups.

The shortening of the pectoralis major (attaching from the chest to the upper arm) could (1) shorten upward reach (see figure 5.2 on page 48) and (2) limit the backswing of the arm, reducing the time and distance over which force could be developed. The shortening of the iliopsoas muscles could pull the back into a sway, reducing the springlike function of the curves of the vertebral column and decreasing its ability to absorb the many shocks received by the body each time the person hits the ground in landing or running. The shortened iliopsoas could also increase the shock trauma to the small of the back (see figure 5.6).

Can you see how tight quadriceps muscles (found on the front of the thigh and attaching below the knee) could limit the preparatory backswing of the leg, reducing the time and distance over which force could be developed? (See figure 4.7 on page 33.) Can you also see how tight hamstrings (found on the back of the thigh and attaching below the knee) could limit the forward swing of the leg, reducing the time and distance over which force could be developed?

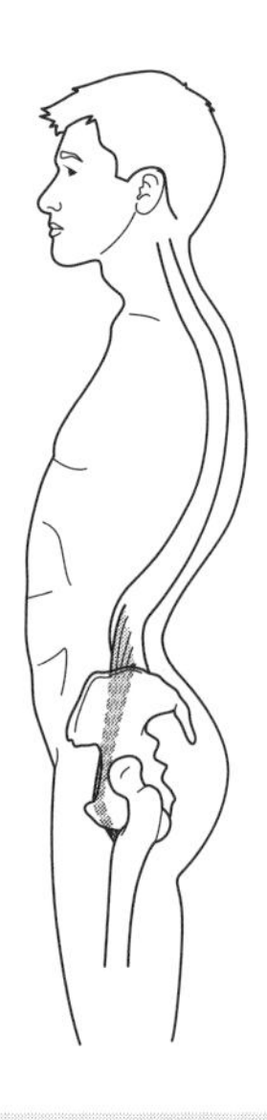

FIGURE 5.6 Find the iliopsoas muscles in this figure. Can you see why a tight iliopsoas would lead to back problems?

To find a good program of strengthening and stretching, discuss your needs with a physical educator or a physical fitness specialist. You should also learn the location and attachments of the involved muscles as well as the most current exercise information. This will enable you to carefully and continuously evaluate the correctness of your exercise technique. Incorrect exercising can be harmful and can work against your efforts toward improvement.

Because connective tissue shortens if not periodically stretched, a heavily used muscle may actually shorten, causing tightness and a loss in range of motion.

Activities

Specific exercises for each of the alignment problems discussed in this chapter are readily available elsewhere. However, I have found that information on two areas seems to be limited or erroneously covered in popular sources. The first area is the lengthening of the iliopsoas, and the second is toe touching.

Lengthening of the Iliopsoas

The stretching of the iliopsoas is probably the most neglected area of possible difficulty. Few people have ever heard of this muscle, although it contributes to countless backaches every year.

The iliopsoas is attached to the inside of the vertebral column in the region of the lower back (lumbar area), passes over the front of the bony pelvis, and then runs downward and attaches to the inside of the upper leg bone (femur). Therefore, the iliopsoas is somewhat like the tight strings of a cello (figure 5.7). When this muscle shortens, it pulls the small of the back forward and down, leading to a sway in the back and increasing the pressure on the vertebral column. (Because the vertebral column is such an important part of the body, a medical evaluation should be completed before a person performs exercise that involves this area.)

Here are three exercises that can be used to help keep the iliopsoas muscles lengthened:

- *Flattening the small of your back while lying flat:* Try to keep your thighs (iliopsoas attachment) on the floor while attempting to hold your lower back on the floor. The flattening of the back is accomplished by contracting the abdominal muscles (figure 5.8). The ability to execute this exercise with ease is used as an evaluation of the length of the iliopsoas.

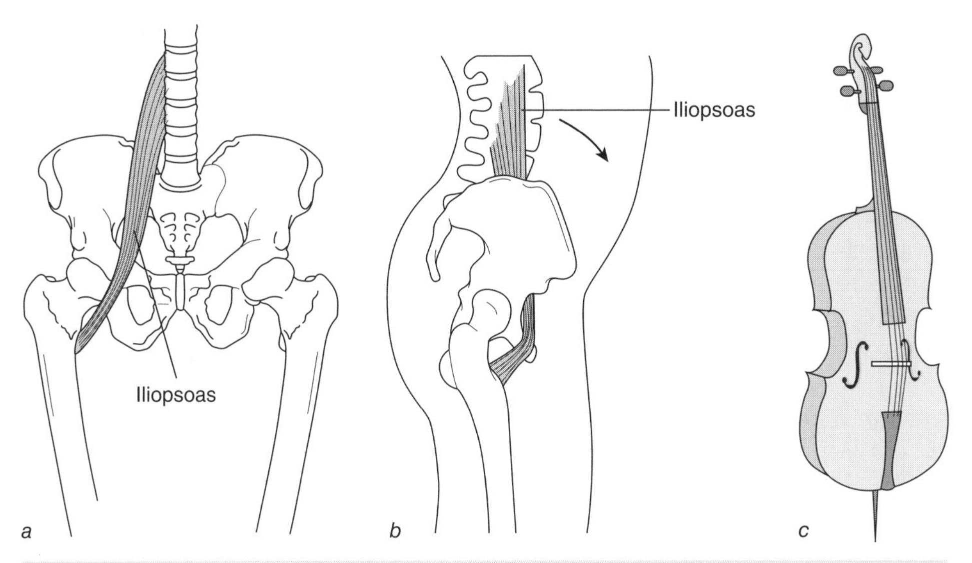

FIGURE 5.7 *(a)* A tight iliopsoas needs to be lengthened; *(b)* side view of the iliopsoas muscles; *(c)* a tight iliopsoas can be compared to the tightness of cello strings.

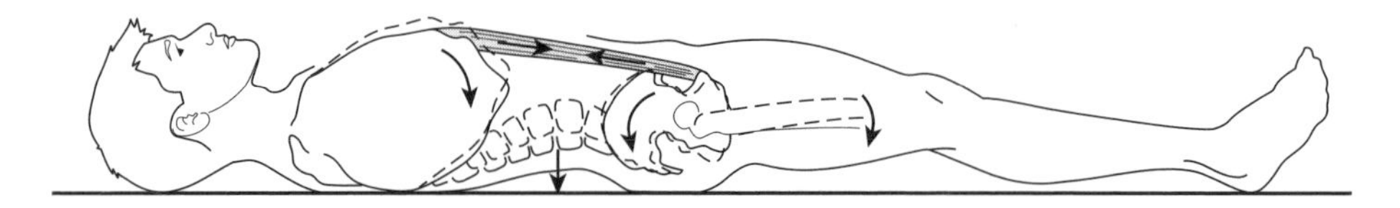

FIGURE 5.8 When lying down, can you touch the small of your back to the floor? If not, you may have a tight iliopsoas.

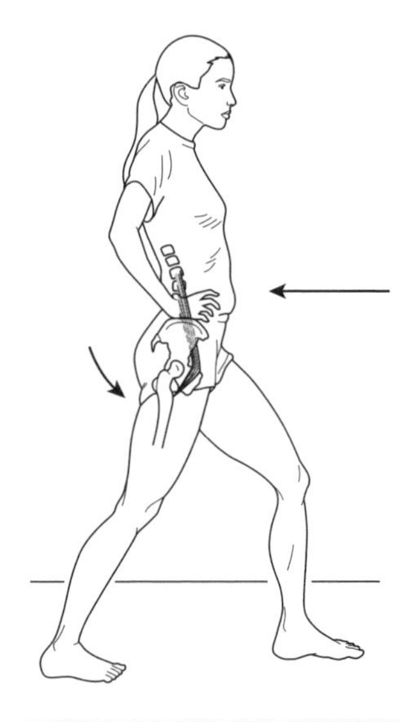

FIGURE 5.9 The fencer's stretch is one way to stretch and lengthen the iliopsoas.

- *Bringing your knee to your chest while lying flat:* While lying flat on your back, bring one knee to your chest (hugging it) while you keep your other leg and the small of your back touching the floor. Hold this position for 30 seconds and repeat 5 times. Then repeat the exercise with the other leg. Do this at least twice a day. Note that the leg that is down is on the side that you are actually stretching and should receive your attention.
- *The fencer's stretch:* Stand and place your feet in a stride position, both feet flat on the floor (figure 5.9). The position of the back foot holds the lower attachment of the stretched iliopsoas in place. The toes of the back foot should point straight ahead. Rotate the pelvis under you, pushing the upper pelvis (waist) backward while the lower pelvis rotates and presses against the iliopsoas. By pushing the upper attachment (small of the back) backward, you are further stretching the iliopsoas on the back-foot side. Hold this stretch for 30 seconds. Change your stride and repeat for the other side. Do this 5 times on each side. Try to perform this exercise at least twice a day if you tend to have tight iliopsoas.

Toe Touching

Toe touching has long been a standard exercise for maintaining the length of the muscles in the back of the upper leg (hamstrings). Two recommendations have been made for this exercise: (1) Do not include a bounce (you do not want to evoke the stretch reflex or damage tissues). The toe touch should be done slowly and held at the point of full stretch. (2) Do not let the knees hyperextend when toe touching, because this position may increase the potential hazard to the structures of the knee. To prevent hyperextension and excessive pressure within the knee joint that can occur in the standing position, exercisers have been encouraged to do their toe touches from a straight-leg, sitting position or while lying on the back. When lying on the back, the person pulls one leg toward the ceiling while the other leg remains flat on the floor. All stretching should be done slowly and should be held for several seconds.

Remember that your body plays an important role in your involvement in life. Treat it well—in addition to exercise, make activities such as Skates and Rag or Rug Hockey (see chapter 10) a lifetime sport!

True or False Review Questions

Information about each question is found on the page number provided within the parentheses following the question.

REMEMBER THAT A PARTIALLY FALSE QUESTION IS CONSIDERED FALSE.

Answers to the questions and additional information are found in appendix C.

1. Researchers have found that people may need to supplement their activity program with some additional physical preparations, such as strengthening the stabilizers of the pelvic and shoulder girdles, developing and maintaining muscle tone in the antigravity or postural muscles, and stretching muscles that tend to tighten (page 45).
2. Forward shoulders could be caused by tight upper back muscles or weak chest muscles or both (pages 45, 49-50).
3. A swayback could result from tight iliopsoas muscles or weak abdominal muscles or both (pages 46, 50; also see figure 5.6 on page 50).
4. We need strength, endurance, and muscle tone for stabilization (page 47).
5. We need muscle length for range of motion in order to increase the time and distance over which force can be developed (pages 48, 50; also see figure 5.3 on page 48).
6. When a muscle has good muscle tone, a number of muscle fibers are continuously in a state of contraction. The many fibers within the muscle share the task, rotating which fibers are in contraction at any one time. This prevents fatigue and allows for a state of continuous readiness, which makes it possible to make immediate adjustments and increases body control (pages 46-47).
7. Weak upper back muscles do not affect a forceful throw, but tight chest muscles could reduce a basketball player's reach for a rebound (page 47; also see figure 5.2 on page 48).
8. People who remain physically active all their lives do not have to consider stretching unless they stop being active (page 48).
9. Range of motion (ROM) can be increased by doing 10 to 20 ballistic or forceful bounce stretches several times a week (pages 48-49).
10. Both tight quadriceps (the muscle on the front of the thigh and attaching below the knee) and tight hamstrings (the muscle on the back of the thigh and attaching below the knee) can reduce the time and distance over which force can be developed (page 50).

PART III

Moving Objects

CHAPTER 6

Rebound, Deflection, Spins, and Roll Patterns

Being able to predict what an object will do as it rebounds, deflects, or rolls is an invaluable skill. This knowledge allows you to be prepared, and it increases your chance of controlling the object so that you can accomplish your goal. Consider how this awareness could improve your ability to predict where you should move to intercept, catch, or hit a bouncing ball or where to place a layup shot against the backboard in basketball. An initial understanding of rebounds, deflection angles, spins, and roll patterns can help you reduce the amount of time and practice needed to learn related skills through trial and error. This predictive ability or inability is frequently the skill that makes success in movement activities possible or impossible. Inexperienced individuals often get frustrated when they are struggling with this ability. A player may leave a game with a deep sense of failure or may be unable to learn and enjoy the pleasures of a good game because of this inability to make early predictions. Knowledge about rebounds, deflection, spins, and roll patterns can help a player not only predict but also control or cause certain reactions.

Rebounds

If there were no confounding force factors, such as spins, the rebound angle would approximate the approach angle (figure 6.1).

Knowledge about rebounds, deflection, spins, and roll patterns can help a player not only predict but also control or cause certain reactions.

If there were no confounding force factors, such as spins, the rebound angle would approximate the approach angle.

This type of rebound rarely occurs, but it does give us a point of comparison for more careful observation and the development of an awareness of rebound patterns. Understanding these patterns can greatly improve our judgment about, effective use of, and response to the various rebounds that occur in play and sports.

Deflection

Rebound is also called deflection. In the previous example (figure 6.1), one of the two things that came together (i.e., the floor) was seemingly unaffected and did not move or change its position or path in space when contacted by the ball. When the weight of two objects coming together is more equal, both objects can be affected and deflected.

A heavier object deflects less than a lighter object. The reverse is also true. A lighter object deflects more than a heavier object. Speed can also affect deflection. In bowling, where more pins are knocked down by other pins than are knocked down by the ball, too fast a roll will send the pins flying, actually reducing the amount of pin-mixing action caused by pins deflecting against other pins. Can you see why a bowling ball has an optimum (best) speed to be most effective? In some games—such as billiards, pool, and bowling—deflection and the skillful use of it are a primary aspect of the activity.

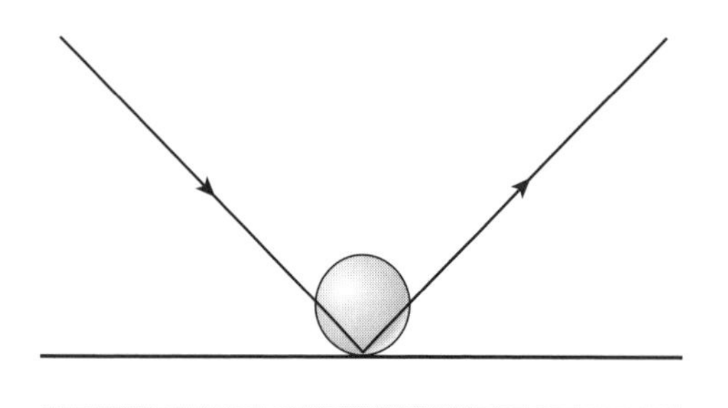

FIGURE 6.1 A friction-free rebound.

Spins and Roll Patterns

A spin is named according to what is occurring on the side of the ball away from the thrower (or kicker or hitter). When viewed from the thrower's position, the far side of the ball may be traveling downward, upward, from left to right, or from right to left:

1. If traveling downward, it is called *topspin.*
2. If traveling upward, it is called *backspin.*
3. If traveling from left to right, it is called *right spin.*
4. If traveling from right to left, it is called *left spin.*

Understanding rebound and deflection without spin establishes the foundation for understanding the complex and exciting possibilities that come about when spin is added. When an object with spin contacts a surface or another object, the spin may give an additional push against the contacted surface, and the object may rebound or deflect in relation to this additional force. Whenever more than one force causes an effect, the resulting combination of forces is known as *resul-*

tant force. We have a resultant force when a spin affects the normal rebound, deflection, or roll pattern.

The only part of the ball that can contribute an additional push force to the ball is the part that comes in contact with a surface (floor, ground, or backboard). Therefore, the direction in which this part of the ball is moving on contact determines the effect.

A ball with topspin (figure 6.2) gives an additional push backward as the bottom of the ball touches the floor. Because of the law of motion stating "to every action there is an equal and opposite reaction" (page 35), this additional push backward tends to send the ball forward. This additional forward force combined with the normal forward rebound causes the ball to travel lower as it comes off the ground. Because both forces are creating forward movement, the ball will have more forward force and will thus go farther.

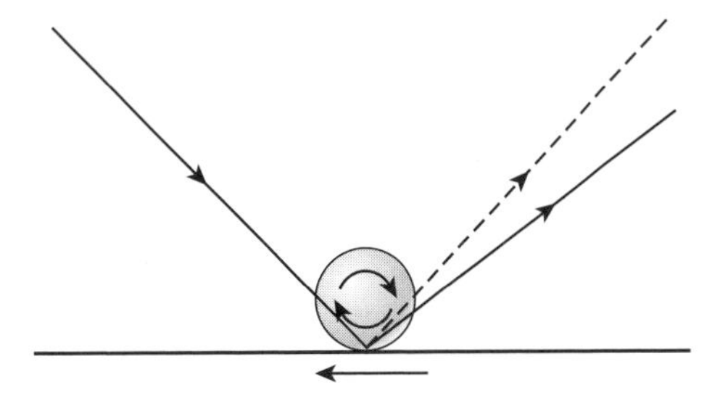

FIGURE 6.2 Topspin. Notice which way the bottom of the ball is pushing against the floor. How will this push affect the ball and why?

A ball with backspin (figure 6.3) gives an additional push *forward* against the contacted surface. The additional reaction is backward, but as long as the forward rebound force is greater than the backward spin force, the ball will still move forward. The resultant force will cause the ball to rebound higher (the attempt to go backward), and because of the two forces moving in opposite directions, the ball will slow down and not go as far.

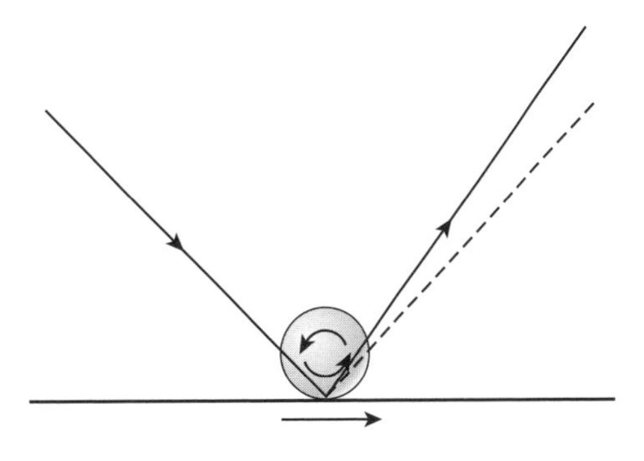

FIGURE 6.3 Backspin. Notice which way the bottom of the ball is pushing against the floor. How will this push affect the ball and why?

The hook ball in bowling is an interesting example of a resultant force caused by spin and forward momentum. The hand position in the ball (figure 6.4), which results in the fingers leaving the ball after the thumb, causes an additional pressure or force on the ball. This adds an over-the-top spin (over the top of the ball) to the forward roll force.

As the forward roll force diminishes later in the roll, the over-the-top spin force causes the ball to cut into the pins (figure 6.5).

Consider a left- and right-handed person's hook ball roll (figure 6.5). Can you determine which way the spin force of each

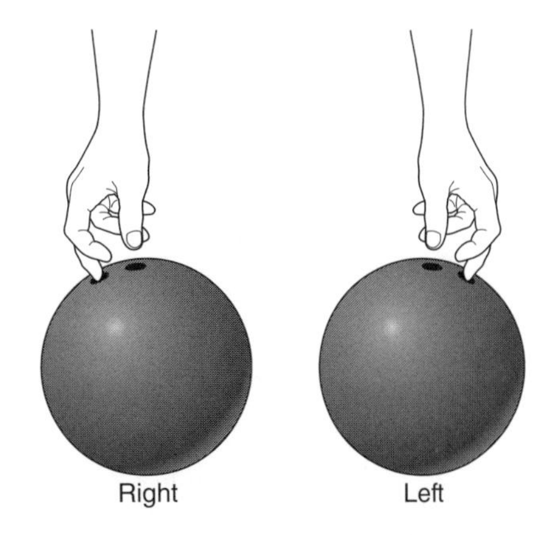

FIGURE 6.4 On a hook ball, the fingers leave the ball after the thumb, thus applying an over-the-top force.

bowler's ball would be pushing against the floor and in turn which way this would cause each bowler's ball to roll (figure 6.6)?

Can you see how this might allow a hook ball bowler to bring the ball into the pocket to involve the center pin, or kingpin (the number 5 pin found in the middle of the pins), in order to cause a better pin action? Because a spinning ball will also add spin to the pins that it hits, the spin of a hook ball will add more pin mix.

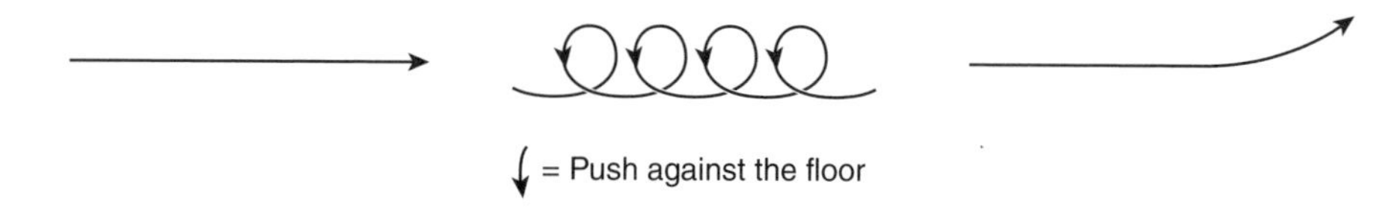

FIGURE 6.5 As the hook ball slows down when it nears the pins, the over-the-top spin causes the ball to move into the 1-3 pocket for the right-handed bowler or into the 1-2 pocket for the left-handed bowler.

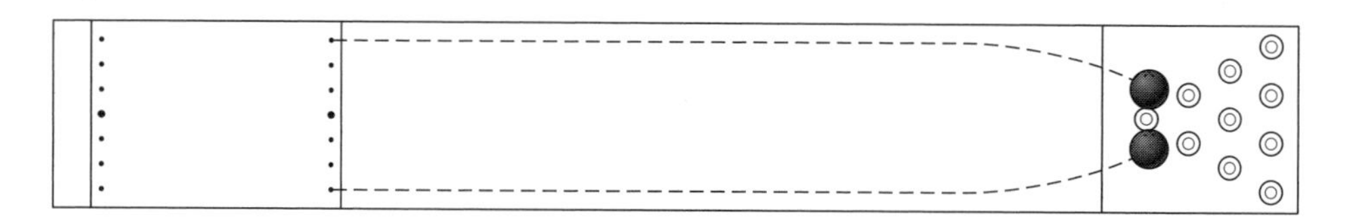

FIGURE 6.6 In a right-handed hook ball, the bottom of the ball is moving or pushing from left to right, causing the ball (as it slows down) to go to the left. For a left-handed bowler, the opposite is true.

Here are some ways that we can use these resultant forces and their effects to our advantage:

1. We can increase our control over the movement of an object so that we have a broader scope of alternatives.
2. We can increase the ways that we can affect other objects contacted by a spinning object.
3. We can confuse our opponents and make it more difficult for them to predict and thus respond effectively.

Although the use of spins requires a greater degree of skill, it does open up a whole new world of possibilities and challenges.

If backspin is put on a foul shot in basketball and the ball touches the backboard, what direction will this additional spin tend to take the ball? Is this advantageous? (See figure 6.7.)

Why would adding a left or right spin to a layup shot be helpful? Which spin would you want to use?

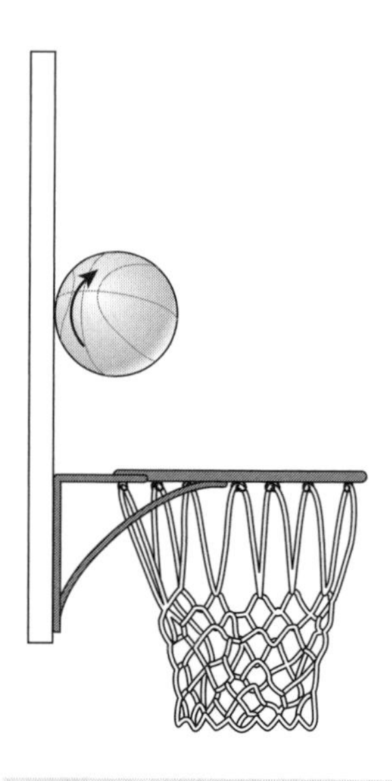

FIGURE 6.7 Is it possible that the friction created by the backspin could also slow the ball down, allowing gravity to pull it downward toward the basket?

Would it be different from the other side of the basket? Would one type of spin reduce the forward force and help the ball drop through the basket? Would the spot where the ball is placed on the backboard need to be different with a spin than if the ball had no spin on it? Paper and pencil may help in answering these questions.

A golfer may want the golf ball to land on the green but not to continue to roll beyond it and into a sand trap. How could the golfer make the ball "bite" and stop (figure 6.8)?

If a tennis player wants the tennis ball to come up low and have additional speed, what type of spin would the player employ (figure 6.9)? If the player's opponent is deep in the court and the player wants the ball to hit and stay in the front part of the court, what type of spin should the player use (figure 6.10)?

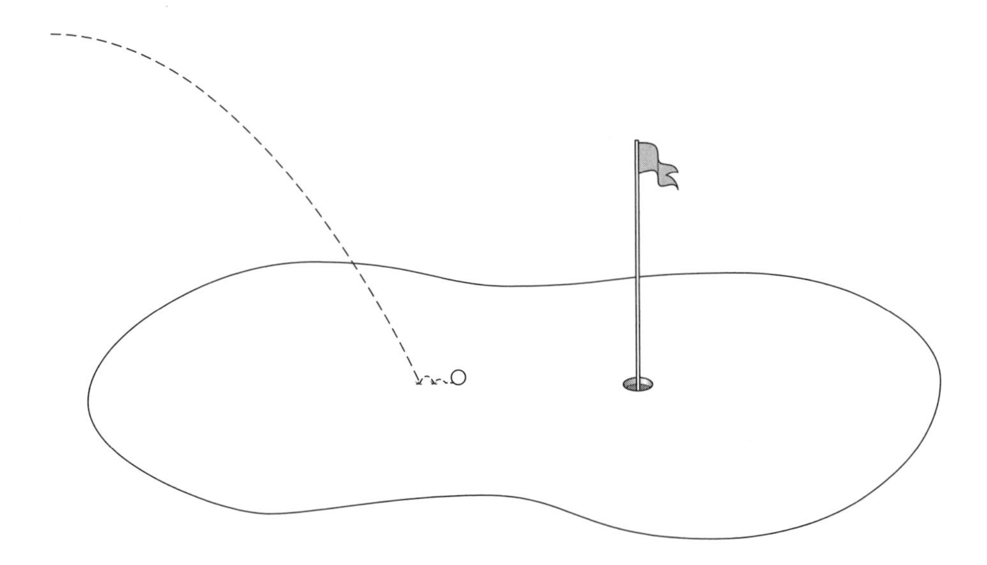

FIGURE 6.8 What type of spin will cause a golf ball to bite when it lands on the green? Why would a golfer want this to happen?

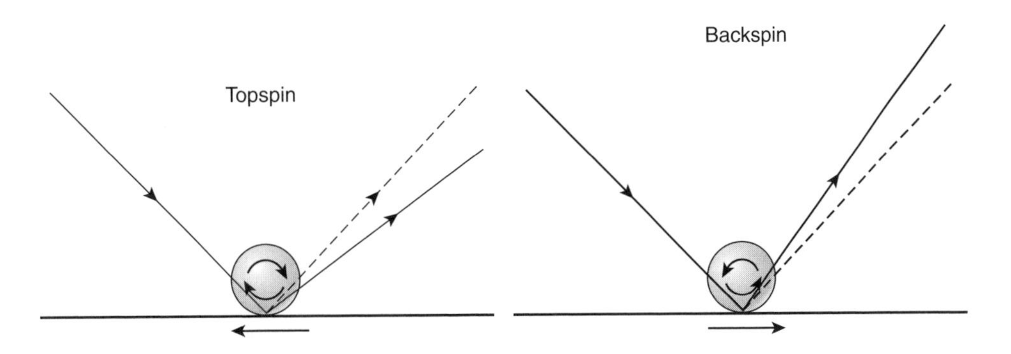

FIGURE 6.9 Why might a player want to use topspin in tennis?

FIGURE 6.10 Why might a player want to use backspin in tennis?

Activities

- Encourage players to play with a variety of balls so they will experience a variety of responses. By doing this, players will begin to be able to predict, read, and adapt to various rebound, deflection, and roll patterns. Observe the players' improvement over several weeks. Their early experience is invaluable and will allow for an almost intuitive response later. (See Racket and Balloon, Blanketball, Rebound Board, and Frantic Ball in chapter 10.)
- See if you can bounce a ball into a box or wastepaper basket by judging the angle. If this is too difficult at first, tilt the box against the wall. Can you bounce the ball at various angles and from varying distances?
- Without any spin, bank a ball off a wall. Hitting the same spot each time, vary your approach angle coming from the left or the right. Can you see how this change affects the rebound angle? Now mark several points along a horizontal line about shoulder high. Hitting each of these, note the angles of rebound.
- Bouncing a ball on the floor, give it a hard backspin as you push it forward and downward. Can you see the effect as it hits the ground? Now do the same with topspin. Alternate left and right over the topspin on the ball and try to determine the effect on the rebound. Remember that the part of the ball that contacts the floor or ground determines the direction of the rebound.
- Throw a ball against a wall using several different types of spin. Can you differentiate the various results and clarify why these occurred? Now consider the effects of left or right spin on a normal rebound angle in a layup in basketball. Can you see how the approach angle might affect the rebound angle? Can you see how the point where you contact the backboard could affect the rebound angle? Can you see any advantage to adding a spin to affect the rebound angle? Can you begin to determine where you should place the ball on the backboard and what type of spin you should use to make a successful layup from either side of the basket?
- Using a ball, a container, and a wall, try to bank a spinning ball into the container so it does not bounce back out. To become proficient at this, you will have to control the rebound angle and be able to "kill" the momentum of the ball. Try this experiment from both the left and right side. Note how you must reverse the spin. Can you see any relationship between this experiment and a layup shot? Note that the same type of spin that produces a rebound that directs the ball more out toward the basket also helps reduce the momentum of the ball so that it tends to drop toward the opening of the hoop.
- Find a pool or billiards table and perform these activities: (1) Without using spin, bank the cue ball off the sides of the table. Note the resulting rebound angles. Can you begin to predict the path of the ball? As you get better at predicting the rebound angles, put a target ball or coin on the table. Bank the cue ball in such a way that it will rebound and hit the target ball or roll over the coin. (2) Roll the cue ball into another ball. Repeat, hitting the ball at various points on it. Can you begin to predict roll patterns? (3) Now put a sidespin on the ball that you are rolling by spinning it like a top from above. Observe how this changes

the roll pattern and the effect it has on the other balls it hits. Can you plot the difference between the patterns in activity 2 and 3? (4) Spin the rolled ball in the other direction. Can you plot and compare the patterns in activities 2, 3, and 4? Can you draw any conclusions?

- Set up two obstacles in line with each other (one behind the other). Can you roll a ball with spin so that you miss the front object and contact the object behind it? Can you do this with both a left and right spin?

Players need experience with rebounding, deflection, roll patterns, and spins in order to respond effectively to activities that require this skill. Activities such as those just described, bowling, and billiards give players the repetitive experience needed with multiple opportunities to view the way a ball moves under various conditions. All participants should be encouraged to play games that contribute to their basic skill abilities (see chapter 10). Success in responding to any moving object is based on either conscious or intuitive awareness. This can best be gained by a combination of thinking and experiencing.

Observe various movement activities. Can you see any use of rebound, deflection, and roll patterns?

True or False Review Questions

Information about each question is found on the page number provided within the parentheses following the question.

REMEMBER THAT A PARTIALLY FALSE QUESTION IS CONSIDERED FALSE.

Answers to the questions and additional information are found in appendix C.

1. In bowling, more pins are knocked down by the deflection of other pins than are knocked down by the ball (page 58).
2. The faster a bowling ball is traveling, the more pins it will knock down (page 58).
3. When an object with spin contacts a surface (such as the floor, ground, or basketball backboard), the spin may give an additional push against the contacted surface, and the object will then have a resultant force and will respond to this combination of forces (pages 58-59).
4. How the bottom of a spinning ball is pushing against the floor determines how the spin will affect the roll or rebound (pages 58-60).
5. A ball with backspin will rebound lower and slower than a ball with topspin (page 59).
6. If a basketball foul shot has topspin on it and it hits the backboard, this will help push the ball downward toward the basket (page 60; also see figure 6.7 on page 60).
7. When a basketball player is performing a layup from the left side of the basket, a right spin would slow the ball down and, if placed correctly (at the best rebound point on the board in relation to the resultant force), could help drop the ball into the basket (pages 58, 61, 62).
8. Backspin (far side of the ball moving upward) on a golf ball will help it to “bite” as it hits the green so that it does not roll off the green into a sand trap (pages 58, 59, 61).

9. A left-handed hook ball in bowling will travel to the left as it slows down. Think about which way the bottom of the ball will be pushing against the floor (page 60).
10. Knowledge of the effects of various spins can be helpful in making decisions in several sports (pages 57, 58-61).

CHAPTER 7

Projectiles

People who are unaware of the mechanical principles of projectiles must learn certain skills through the arduous and frustrating process of trial and error. Many never discover the seemingly mystical flight patterns that can be effectively used when a person understands the concepts of the parabolic path, spins, the effects of air pressure, and the role that the center of gravity plays when a person becomes a projectile.

Principles of Projectiles

Knowledge of the principles of projectiles will enable you to both execute different flight patterns strategically by creating specific results to baffle an opponent and will also allow you to read and effectively respond to various flight patterns. Let's look at some of these principles.

- *The flight pattern of an object is normally fairly consistent. Unless modified by wind, spin, or some characteristic of the object that affects air pressure in a specific way, the pattern is parabolic* (figure 7.1).

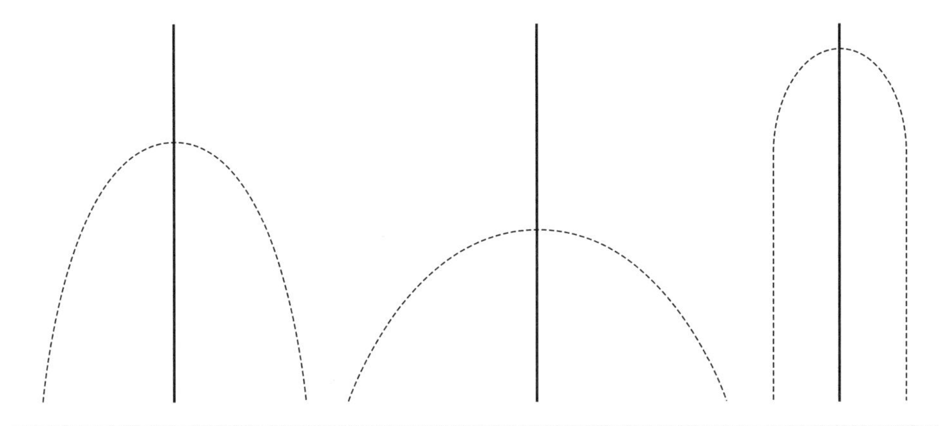

FIGURE 7.1 Three parabolic flight patterns. Each parabolic flight mirrors the first half of its flight in reverse.

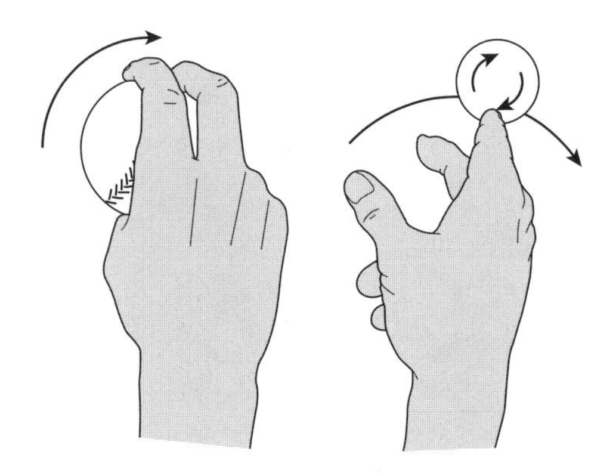

FIGURE 7.2 A ball can be made to spin by putting an off-center pressure on it, which will cause it to rotate.

This simply means that the second half of the flight mirrors, or duplicates, the first half of the pattern in reverse. Being aware of this may help those receiving an object to determine the path it will take and where it will come down. Thus, they become more able to read the flight of an object.

The exception to the parabolic flight principle that we experience most frequently in sports involves a spinning object. We can affect the normal flight pattern by creating an off-center force on the projected object, thus causing the object to rotate. This in turn creates an uneven air pressure system around the outside of the object (figure 7.2).

- *A spinning object will pull air molecules around its surface as it spins.*

The exception to the parabolic flight principle that we experience most frequently in sports involves a spinning object.

When the air molecules traveling around the surface of a spinning object collide with the air molecules into which the object is traveling, a high-pressure area forms. On the opposite side of the same spinning object, the air molecules are being dumped in the other direction, creating a low-pressure area (figure 7.3).

- *An object will tend to be pulled by a low-pressure area (suction) and pushed by a high-pressure area, thus causing the object to move in the direction of a low-pressure area.*

This applies to all spins, whether they be left, right, top, back, or diagonal spins. Figure 7.3 shows a ball with backspin. This ball will tend to lift as it travels through the air because of the high-pressure collision area that will form at the front bottom of the ball; a low-pressure (suction) area forms as the air at the upper front of the ball is pulled from that area.

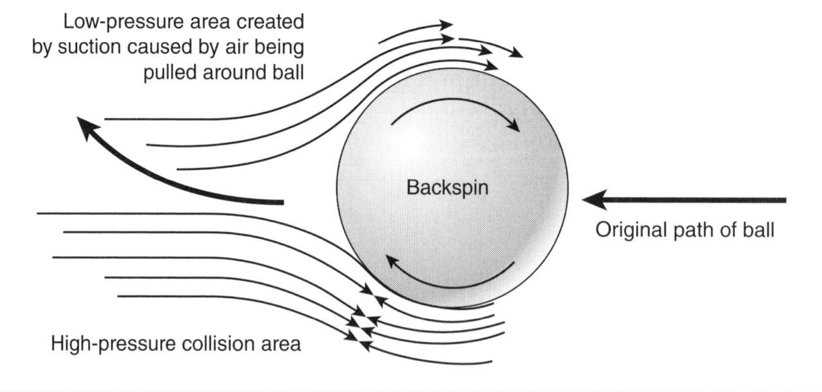

FIGURE 7.3 A spinning object pulls air molecules around its surface as it spins. This creates high- and low-pressure areas, causing the object to move in the direction of the low-pressure area.

The various types of spins will have the following effects on the path of a round object:

1. In a topspin, the ball will go down as the air molecules collide on the top front of the object.
2. In a backspin, the ball will go up as the air molecules collide on the bottom front of the object.
3. In a left spin, the ball will go left as the air molecules collide on the right side of the object.
4. In a right spin, the ball will go right as the air molecules collide on the left side of the object.
5. In a diagonal spin that mixes topspin and right spin, the ball will go down and right as the air molecules collide on the top and left side of the object.
6. In a diagonal spin that mixes backspin and left spin, the ball will go up and left as the air molecules collide on the bottom and right side of the object.

When the air molecules traveling around the surface of a spinning object collide with the air molecules into which the object is traveling, a high-pressure area forms. On the opposite side of the same spinning object, the air molecules are being dumped in the other direction, creating a low-pressure area.

With each of these spins, a low-pressure area is formed on the opposite side of the spinning object as the air molecules traveling around the object are being pulled in the same direction that the object is traveling (see figure 7.3). A low-pressure area pulls the object in the same direction that the high-pressure area pushes it.

Could you mix other combinations and determine where the ball will go? If you wanted the ball to dive, would you be able to determine what type of spin to apply? Can you think of three specific situations in which spins would be advantageous?

The surface of the object can also affect how much air is pulled around the object. A golf ball has dimples. This provides little pockets that carry additional air to the back of the ball, helping to prevent a vacuum from forming, which might reduce the forward momentum of the ball and cause it to move in an unpredictable flight pattern.

Can you see why a new fuzzy tennis ball may have more spin effect than a balding one? Can you see why a major-league umpire examines a foul tipped ball so carefully before it is allowed to be returned to play?

How to Apply Spin to an Object

A person can make an object rotate by applying an off-center force to it. Off center refers to a force that is not applied directly through the center of an object (see figure 7.4).

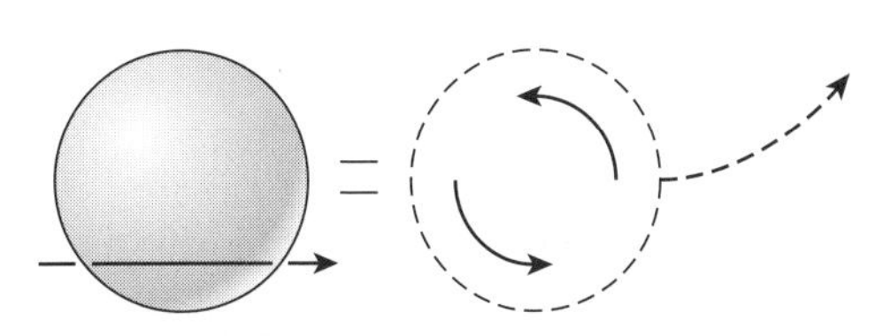

FIGURE 7.4 A ball that has an off-center pressure put on it will spin, causing it to rotate and move toward the low-pressure area created by this rotation. Can you draw in the low and high pressure areas (see figure 7.3)?

This can be accomplished by striking the object off center (as in tennis, golf, or volleyball), by using holes and finger length (as in a hook delivery in bowling), or by using seams (as with a curveball pitch in baseball) to put additional force on one side of the ball. (See chapter 6 for examples of other situations where spin can be helpful.)

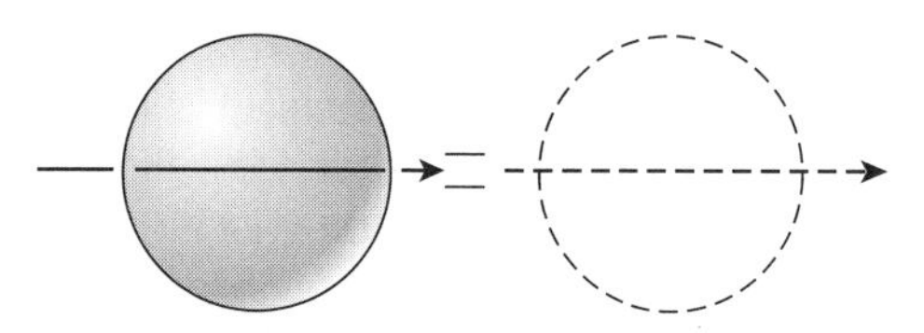

FIGURE 7.5 There are times when a player should hit through the center of an object to get the greatest amount of force or a particular desired direction, such as in a line drive or a home run.

Keep in mind that there are situations in which a mishap can cause an unwanted spin, such as the slice or hook in golf. These can be eliminated by correcting whatever is causing the object to spin. Use the "why, why, why" approach discussed earlier in the book (pages 4-5).

Perhaps you have already heard the admonition "Hit *through* the *center* of the ball" or "Draw your racket *through* the ball." (See figure 7.5.)

The Human Body as a Projectile

The body becomes a projectile whenever it leaves the ground. And our center of gravity does follow a parabolic path. But because we can change the position of the center of gravity within the body, we can do some interesting things to affect our body position during flight and landing (figure 7.6).

- Once your body leaves the ground, the path of the center of gravity cannot be changed.
- You can, however, modify your body parts around your center of gravity while you are in the air. This can raise or lower your body's position in the air.
- When you change the position of any body part, the center of gravity shifts within the body to the new center of weight distribution. However, the parabolic path of the center of gravity does not change.

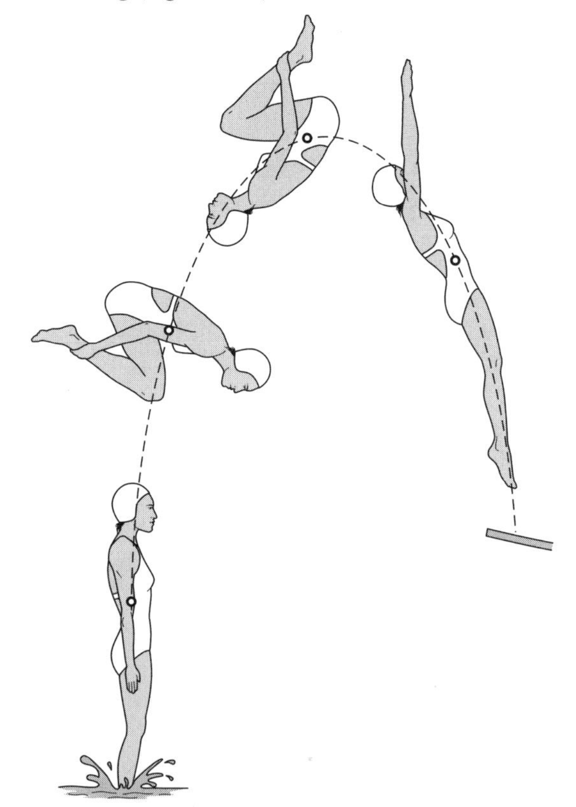

FIGURE 7.6 Note that the center of gravity follows a parabolic path.

Before you can put these ideas together for effective use, you must understand each of these concepts clearly and separately.

When a person is in a normal standing position, the center of gravity is located somewhere behind the umbilicus (belly button). Each individual's center of gravity would be slightly different depending on how his weight is distributed. This will also change with each position in which a person places his body. If a person raises his arms over his head, his weight distribution has changed. The point that balances the weight evenly must also shift upward. Thus, raising the arms also raises the center of gravity within the body (figure 7.7).

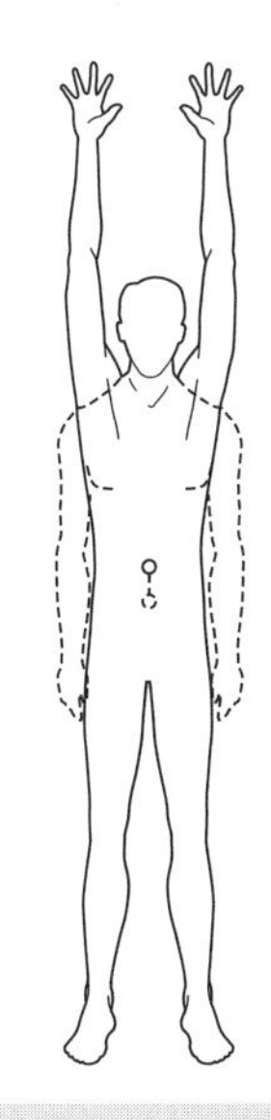

FIGURE 7.7 When a body part is moved, the center of gravity within the body also shifts in that direction.

If a person lifts an arm or leg to the side, the center of gravity will shift in *that* direction in order to evenly divide the new weight distribution. Remember, *any positional change of the body relocates the center of gravity within the body.* But keep in mind that the parabolic path of the center of gravity cannot be altered once the body has left the ground and has become a projectile.

Before applying the combination of these principles to a human body in flight, let's review the last concept.

If you reposition your weight in any direction by moving a body part (i.e., moving arms, legs, or head or bending the trunk), your center of gravity will also shift *in that direction* within the body to divide the newly distributed weight around the center of gravity. For instance, if you lift your leg to the right, your center of gravity will shift up and to the right within your body to become your new balanced center of weight distribution. Although you can change the position of the center of gravity within your body during flight by changing the position of your body parts, you cannot change the flight path of the center of gravity once your body is airborne.

So, the body position in the air will change in relation to the new weight distribution, while the center of gravity of the body retains its original parabolic flight pattern.

If your center of gravity shifts upward in your body because you raise your arms, but the path of your center of gravity cannot change, then the position of your feet in the air will be lower than before you raised your arms (figure 7.8).

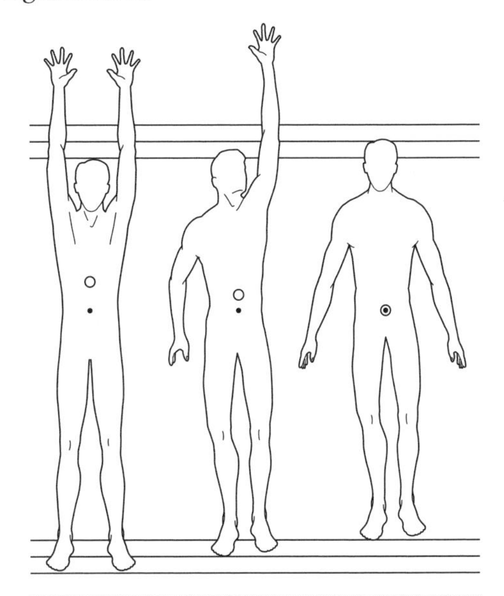

FIGURE 7.8 Note how the body changes levels or heights as the body parts are moved. Which change raises the body the most? The ○ symbol is the center of gravity.

The opposite also holds true. If you lower your arms during flight, the position of your feet should be higher than if you did not lower your arms. Timing will be important in using this factor effectively.

If you lower your arms during flight, the position of your feet should be higher than if you did not lower your arms.

This poses some interesting questions:

- Should a person raise one or both arms in a jump and reach activity?

- When a person is going for height, should both legs be in a downward position at the height of the jump (layup, dunk shot, spike in volleyball)? (See figure 7.9.)

- Considering that the parabolic path of the center of gravity does not change once the body is in flight, could a person put his feet farther *forward* in a broad jump by having his arms back just *before* landing (causing the person's center of gravity to be farther backward in his body and causing his legs to be farther forward than if both his arms and legs were forward)? The person would still need to bring his arms forward on landing in order to bring his center of gravity forward so that he would not lose balance and fall backward.

- At the highest point of a dive, would a person be able to get the appearance of more height—or "hanging" in the air—if she piked her dive (figure 7.10)?

FIGURE 7.9 How might a volleyball player get more height to spike the ball? Should both arms be up?

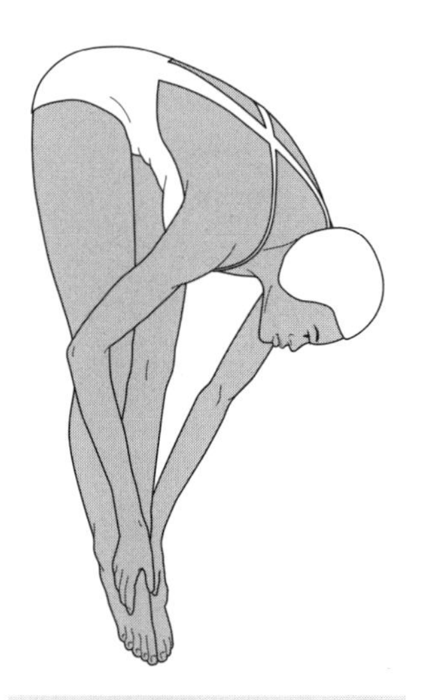

FIGURE 7.10 What will happen to the position of the center of gravity in the body when the arms and legs are brought down into the pike position? Because the center of gravity changes within the body, but the path of the center of gravity in the air does not change (see figure 7.8), what will the body appear to do in the air?

These same principles also become important if you are attempting to clear an obstacle—such as a hurdle, a high jump bar, or a pole vault bar—or if you are vaulting some object. Perhaps a historical example of the changing high jump technique would help make this concept clearer. Over the years, three basic and distinct techniques have evolved: scissors, straddle style, and Fosbury flop (figure 7.11). Each has brought new height records.

Now let's approximate the center of gravity in relation to each of these shapes (see figure 7.12).

Now let's draw in high jump bars at the height at which they can be cleared when using each of the specific high jump techniques (using the same center of gravity path) (figure 7.13).

Scissors: In the scissors technique, the high position of the upper trunk plus the lifting of both legs raises the center of gravity fairly high in the body. This places a great deal of body mass between the center of gravity and the bar.

Straddle style: In straddle style, the body is laid out horizontally, putting the center of gravity lower in the body which raises the body and reduces the amount of body mass between the center of gravity and the bar.

Fosbury flop: In the Fosbury flop technique, the center of gravity is actually outside the body. The flop takes the upper trunk over first. As the upper body clears the bar, it is quickly lowered (lowering the center of gravity within the body), and you can almost see the lift of the body.

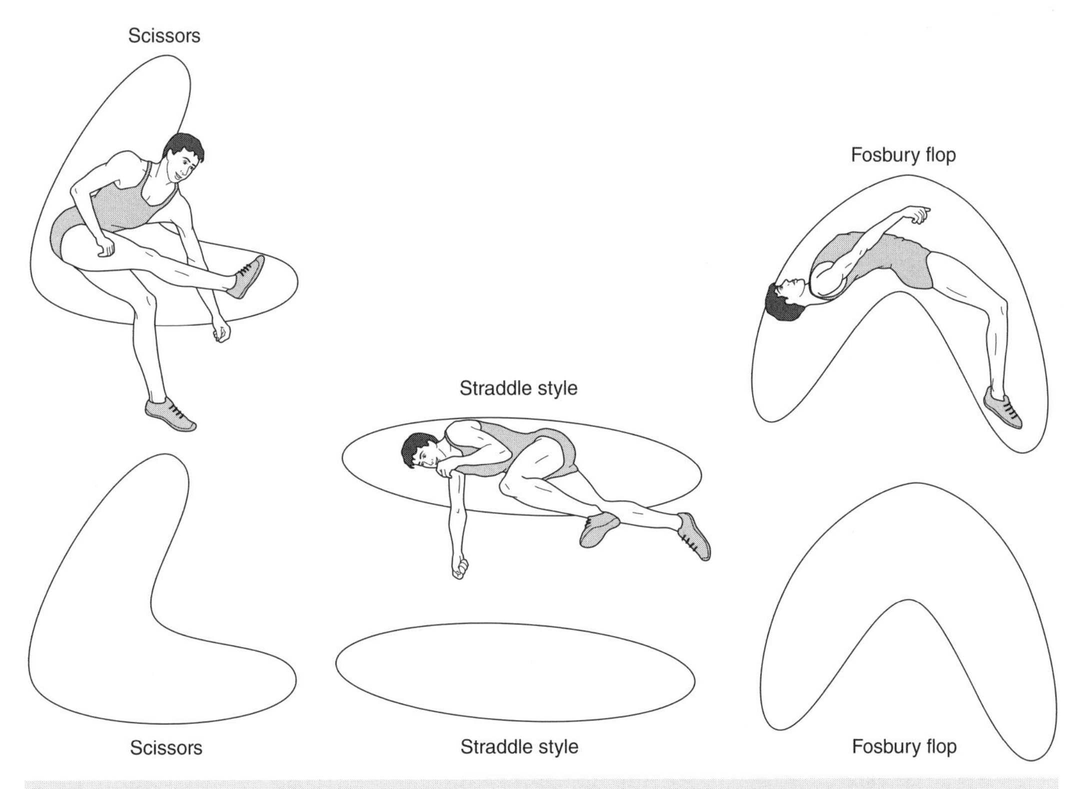

FIGURE 7.11 Three high jump techniques have evolved over time. Which technique will allow the jumper to clear the bar at the highest point?

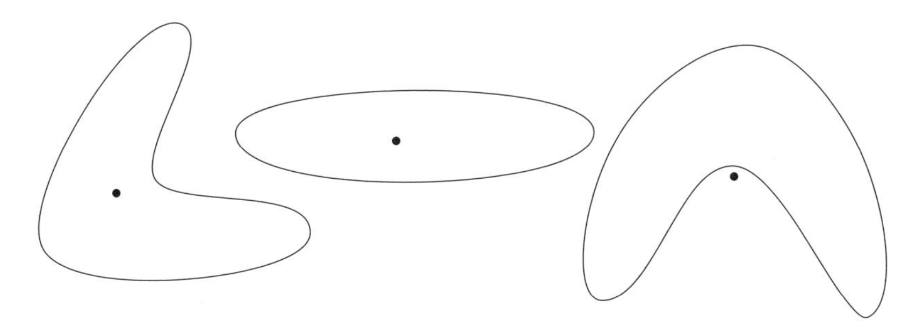

FIGURE 7.12 As the body positions changed, the position of the center of gravity (•) within the body also changed.

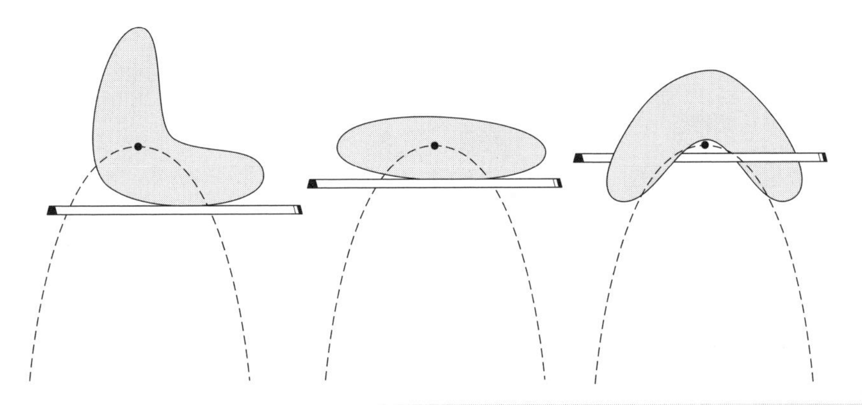

FIGURE 7.13 Note how these changes allowed the body to pass over the high jump bar at different heights.

Can you see how these evolving techniques have used the concept of positioning the body while in flight to shift the position of the center of gravity within the body (thus decreasing the possibility of the body interfering with the bar)?

- Newton's law of action and reaction: "To every action there is an equal and opposite reaction" will also apply to changing the position of the body in flight. If one body part is moved, a counterforce will be exerted on another body part. If an action occurs above the center of gravity, it should produce a reaction below the center of gravity.

Thus, if an arm or the head is dropped, another body part may rise. In the Fosbury flop, the head and shoulders are dropped downward after passing over the bar. This lifts the legs, helping them to clear the bar. This general concept can be applied to various positional changes in the body while in flight.

Of course, the technique used in each of the high jumps discussed becomes progressively more sophisticated, and for participants without skill, the risk of injury also increases. For this reason, beginners should start with the simpler techniques. Can you use information such as this, and transfer it to various situations?

In hurdling, should an athlete try to lean his trunk forward over the hurdle rather than leave his trunk in a more upright position (figure 7.14)? Why or why not? (Relate this to the center of gravity and body position and the action–reaction counterforce effect.)

Would it be helpful for a hurdler to bring his lead leg down quickly, thus "stepping" the hurdle (figure 7.15)? Does this have anything to do with the relationship of body position to center of gravity or with the action–reaction counterforce effect?

To help you answer the previous questions, you may want to study the answers to the true or false questions in appendix C (questions 4, 5, and 10 on pages 140-141).

FIGURE 7.14 When the hurdler leans down as the body begins to go over the hurdle, what does this do to the center of gravity within the body? What does this do to the body in relation to the hurdle? Why does a hurdler want to stay low and close to the hurdle when competing?

FIGURE 7.15 If the hurdler brings the foot down just as the leg passes the hurdle, what does this do to the center of gravity within the body and to the body in relation to the hurdle?

Observe sport events in which humans are projectiles. Can you see any situations in which the principles we have been discussing could be applied?

Activities and Observation

- *Jump and reach:* Attempt to touch something above you. Make sure it is safe before you try. First reach with both hands. Then reach with one hand. Then reach with one hand while bringing the other arm down to your side. (For some people, this may feel awkward and distracting at first and may reduce the height of their initial attempts.)

After you become comfortable with these three forms of jumping, you should determine whether there were any measurable differences or improvements in your performance.

- *Jump and reach with various leg positions:* Jump and reach with the "lifting" leg staying in the up position; then jump and reach with the "lifting" leg being brought down just before the height of the jump is reached. (Timing may be difficult.)

Was there any measurable difference?

- *Observe a baseball or softball player attempting to prevent a home run by leaping high in the air to catch a fly ball.* Does this player incorporate the principles related to the human body as a projectile?

Can you devise experiments to test the principles in this chapter? Remember that the center of gravity is also a vital component of balance (see chapter 2). If you are not able to bring your center of gravity over your base of support when landing, you will be faced with the additional problems created by loss of balance. You may want to think about how a person absorbs force (see chapter 4, pages 37-39).

True or False Review Questions

Information about each question is found on the page number provided within the parentheses following the question.

REMEMBER THAT A PARTIALLY FALSE QUESTION IS CONSIDERED FALSE.

Answers to the questions and additional information are found in appendix C.

1. Without spin, wind, or other factors that create various air pressures, an object in flight follows a parabolic path (page 65).
2. A spinning object pulls air molecules around it. When these air molecules collide with the air molecules into which the object is traveling, a high-pressure area forms. At the same time, the air molecules on the opposite side of the object are pulled in the opposite direction, creating a low-pressure area. The object is pulled toward the low-pressure area (pages 66-67).
3. In a diagonal spin that mixes backspin and left spin, the object will go up and to the left (pages 66-67; see figure 7.3 on page 66).
4. When the body leaves the ground and becomes a projectile, the center of gravity follows a parabolic path, but we can change the parabolic path of the center of gravity by changing the position of various body parts (such as raising an arm or leg) (pages 68-70).
5. When a body part is moved, the center of gravity also shifts in that direction within the body, thus changing the body position in relation to the new weight distribution (page 68).
6. When the arms are raised, the center of gravity within the body is raised, as is the body position in the air (pages 68-69).
7. Considering that the parabolic path of the center of gravity in the body cannot be changed once the body is in flight, a person could put his feet farther forward in a broad jump by having his arms back just before landing. This would cause the center of gravity to be farther backward in the body and would cause the legs to be farther forward than if both the arms and legs were forward. Timing would be important in utilizing this effectively (page 70).
8. The high jump technique has changed based on changing the path of the center of gravity (pages 70-72).
9. The high jump technique has changed based on changing the position of the center of gravity within or in relation to the body, thus decreasing the possibility of the body interfering with the bar (pages 70-72).
10. Leaning forward as the body moves over a hurdle and then bringing the lead leg down quickly to take the next step both lowers the center of gravity within the body and raises the body to clear the hurdle (pages 72-73).

CHAPTER 8

Direction and Accuracy

Skill is frequently the combination of the development of a given amount of force and the ability to apply this force in a specific direction. In many tasks, a person needs to understand how to affect direction in order to be successful. This knowledge not only enables the person to determine how and why to do something, but also helps the person find the cause of and solution to undesirable results.

For example, if a flat tennis serve—one that is taken on a full reach and has good force behind it—is going into the net or hitting beyond the service area, this is a directional problem, not a force problem. Using the "curious child" method described earlier (see page 4), we can analyze the results to determine why they are occurring. The contact point on the ball and the angle of the racket face may need to be changed to correct this problem. Modifying the placement of the ball toss may be the way to do this (see figure 8.1 and 8.2).

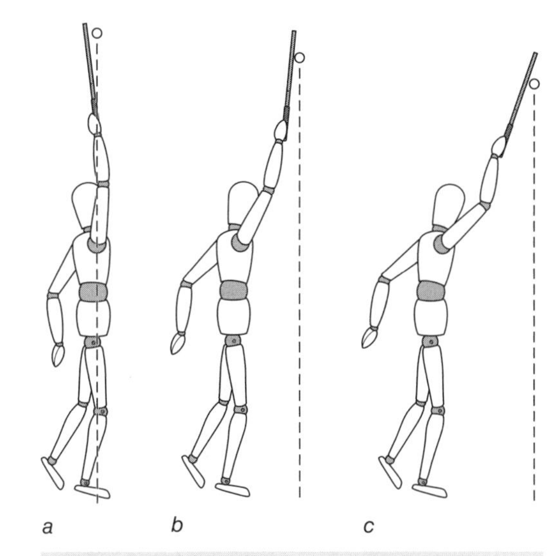

FIGURE 8.1 Can you determine where each of these flat hit serves (*a, b, c*) will go? Notice where the ball toss for the serve is sent up in relation to the body: *(a)* above the server, *(b)* slightly in front of the server, and *(c)* out in front of the server. Does this make a difference? If the serve was frequently going into the net, what would you suggest the server do in relation to the ball toss?

The ball toss should be moved slightly forward if the serve is going beyond the service area, and it should be moved backward if the ball is going into the net. When these changes are made, the contact points and the racket angles become more appropriate, and the serve can be directed more accurately.

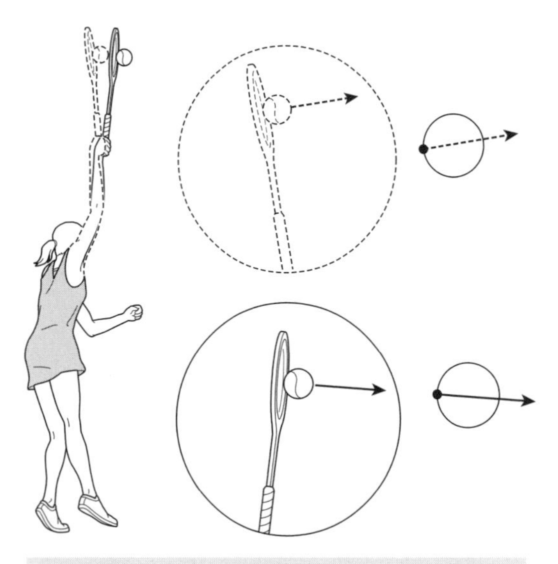

FIGURE 8.2 Can you see how the angle of the racket face changes in relation to where the ball toss for the tennis serve is contacted? If the serve was frequently going beyond the service area, what would you suggest the server do in relation to the ball toss to bring about the best racket face angle? Could this suggestion also be applied to a spin serve? Would changing the position of the ball toss be more consistent than trying to increase or decrease the amount of spin to control the placement of the serve?

FIGURE 8.3 If the ball frequently goes into the net on an underhand volleyball serve, how and why would you have the server move the ball in relation to the server's body?

In the underhand serve in volleyball, the contact point is affected by the position where the ball is held for the serve (figure 8.3).

Can you see how the volleyball serve relates to the previous analysis?

Direction and accuracy are related to the skills of throwing, rebounding, moving, striking, deflection, and so on. Some factors affecting direction and accuracy apply to all of these, while others apply only to some of these skills. As you read and participate, attempt to recognize which factors can be related to which skills. Taking information and making it work for you and others can be challenging. And you must realize that there are no simple answers for all situations but that *you* have to determine all the factors in any given situation and select the most appropriate solutions. As the situation changes—and it will—you need to reevaluate and determine new possibilities.

Factors That Affect Direction and Accuracy

Information and understanding of various factors that affect direction and accuracy should help you select those that relate to the particular situation with which you are dealing.

The Pull of Gravity

Gravity constantly pulls all things toward the earth (figure 8.4). Thus, gravity has a directional effect that we can put to use. By putting just the right amount of force on the basketball or golf ball, we can control its drop through the basket or onto the green. We can hit a Texas leaguer in baseball or softball, execute a drop shot or a well-placed lob in tennis, or cast a plug or fly into the chosen spot in fishing. These skills reflect the ability to compare, contrast, and control the various paths that result from varying amounts of force applied in projecting an object.

Experience allows us to become increasingly intuitive and skillful about receiving and projecting objects. (The games in chapter 10 can help a person gain this experience.) This can be translated into a growing ability to read and control one's environment, which can be important in life situations other than sport and play.

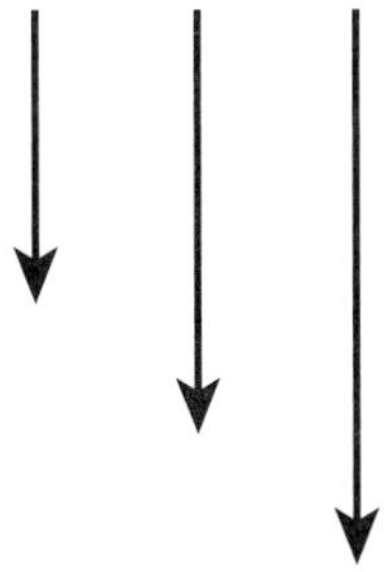

Air Resistance

Air resistance can affect direction in three ways:

FIGURE 8.4 The consistency of the pull of gravity helps us do many things successfully. If we use just the right amount of force in a given task, gravity will do the rest for us.

1. Wind or drafts may be a factor. Sometimes you must reduce their negative effect, and on other occasions, you can use them to your advantage. Have you ever seen a punter who used the wind? Wind can be so helpful that rules are written into track and field regulations prohibiting the establishment of records if the wind direction and speed are too advantageous.

2. Light objects with proportionally large surface areas are affected by air resistance to a much greater extent than heavy objects with a relatively small surface. We see this in the flight patterns of balloons and badminton birds. A badminton bird does not move in a parabolic pattern because the direction of its flight is affected by air resistance (figure 8.5).

3. The various air pressures created by spins can also affect the direction of the flight of an object. (See the section on spins in chapter 7, pages 66-67.)

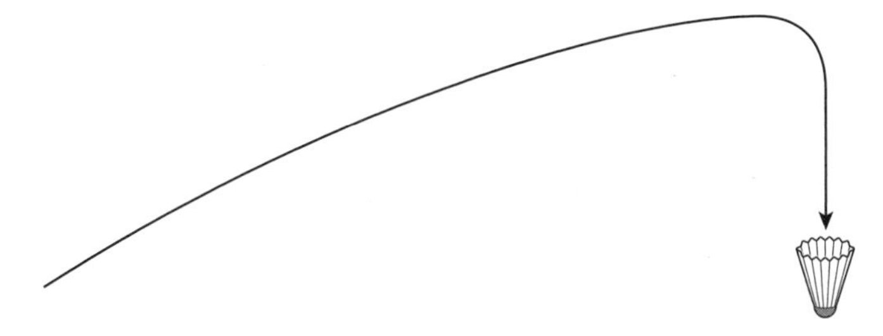

FIGURE 8.5 Because a badminton bird is light and its form creates a great deal of air resistance, its flight path is unusual and may take some getting used to.

Timing and Flattening the Swinging Arc

The term *swinging arc* refers to the path (arc) made by the body part (hand or foot) or striking implement (bat, racket, or club) (figure 8.6). Flattening the swinging arc was dealt with in chapter 4 in relation to force development. Flattening this arc can also help reduce timing problems because a moving object can be contacted along several points in the swing and still go in the desired direction (figure 8.6). Later in this chapter, we'll discuss how it can further improve accuracy.

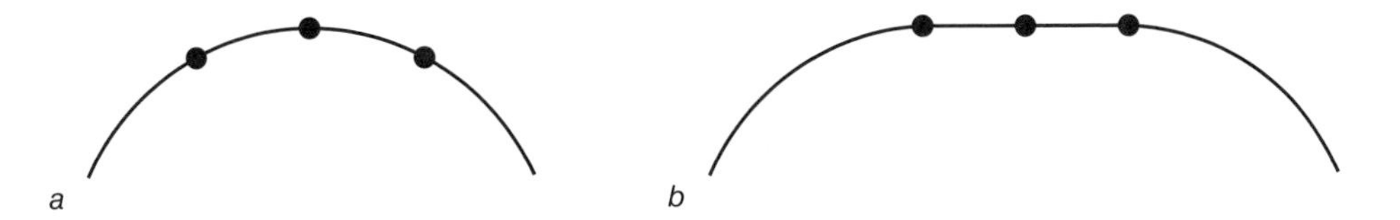

FIGURE 8.6 *(a)* This is what a swing that is not flattened, and *(b)* the same swing that is flattened look like from above.

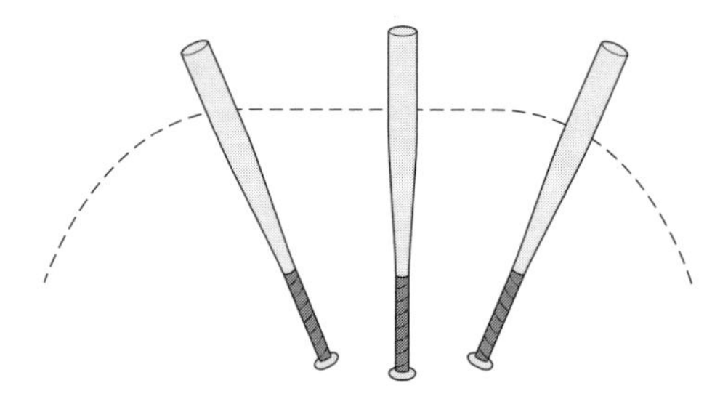

FIGURE 8.7 This is what the flattening of the swinging arc looks like from above in batting.

FIGURE 8.8 Can you see how the transfer of weight from the back foot to the front foot allows the arc to be flattened?

To see the arc pattern, the viewer needs to be parallel to the swing. For a ground stroke in tennis or for batting in baseball or softball, one would have to view it from above (figure 8.7).

In golf, pitching, or a foot pass in soccer, one could best see the arc while facing the player (figure 8.8).

These views place you parallel or horizontal to the movement and give you a better view of the flattening of the swinging arc.

Many inexperienced participants have difficulty with the placement of a throw or hit because they tend to swing in a somewhat circular arc rather than flatten the arc as much as possible. A circular arc creates a constant directional change that makes accuracy more difficult (see figure 8.9). In these circumstances, accuracy relies on the almost perfect timing of the release or impact, which is extremely difficult to achieve, especially for the beginner or less skillful player.

If, however, the arc of the swinging arm, leg, or striking implement can be flattened, the release or impact can occur at any of several points during the swing because the direction of applied force remains the same over an extended period of the swing (figure 8.10).

Can you see how this could be important in golf, pitching, kicking, throwing, batting, racket strokes, a foot pass in soccer, a drive in hockey, and even bowling (figure 8.11)?

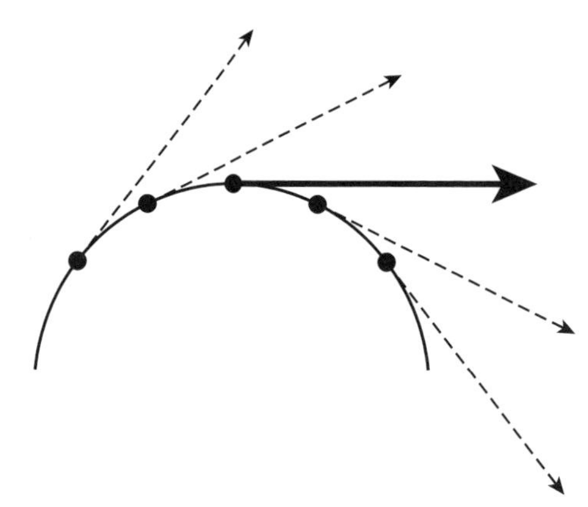

FIGURE 8.9 Because an object will travel in the direction of the applied force, the direction that the object will go changes very quickly when the swing is not flattened. This would make it very difficult to send the object in the desired direction.

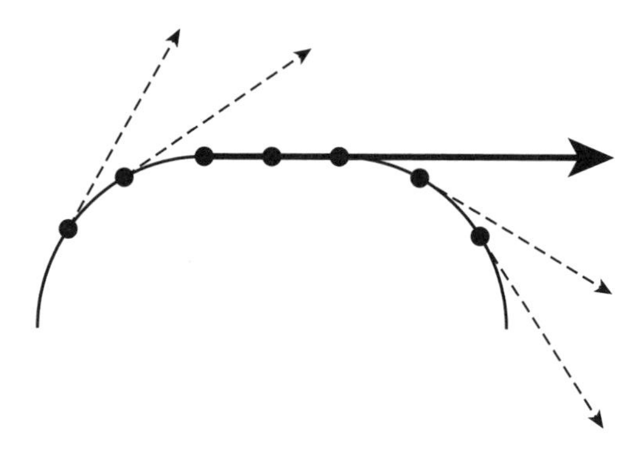

FIGURE 8.10 Can you see how flattening the swinging arc makes it possible to contact the ball at many different points and still have it go in the desired direction? This also allows greater force build up.

The swinging arc can be flattened by using four basic techniques: transferring weight, leading with sequential body parts, moving forward over a bent forward knee, and reaching out during the follow-through. Note that these four basic techniques are also techniques that support force development.

1. *Transferring weight:* In transferring the weight—which can be either side to side or back foot to front foot (figure 8.12)—the person begins swinging around one pivotal point. Then, as the weight is shifted, the person is actually swinging around a second pivotal point, which creates a flattened arc (figure 8.13).

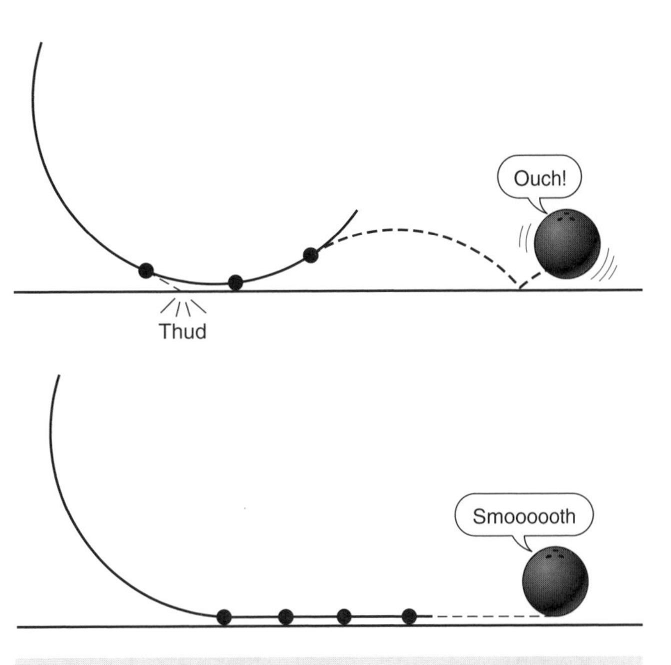

FIGURE 8.11 Flattening the swinging arc also applies to bowling. A smooth delivery increases the accuracy of the delivery.

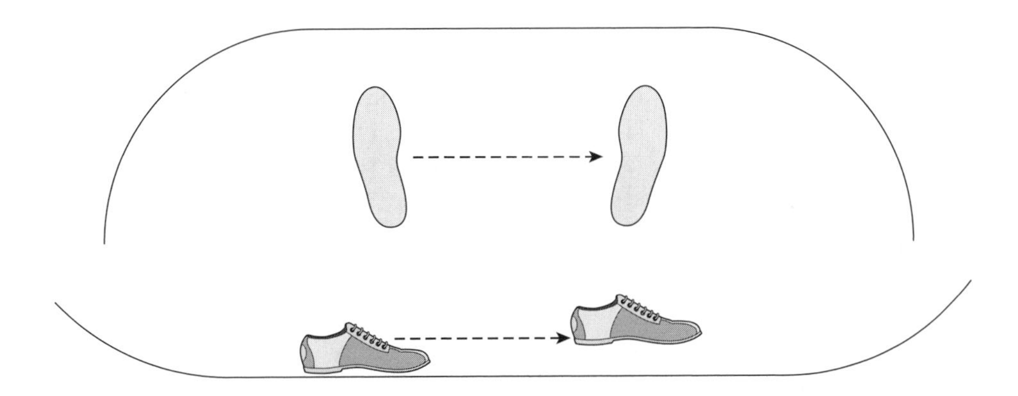

FIGURE 8.12 Weight can be transferred from side to side or from the back foot to the front foot depending on the activity.

2. *Leading with sequential body parts:* Observe a baseball player batting, a person throwing, a tennis player stroking a ground stroke, or a golfer hitting a drive (see figure 4.2 on page 30). Can you see the hips leading, followed by the trunk, then the shoulder? This moves to the next joint and then on to the next. It takes a keen eye and careful observation to see this technique.

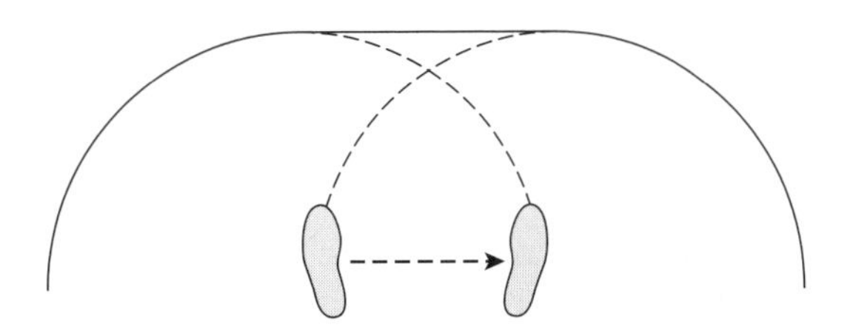

FIGURE 8.13 Can you see how transferring the weight actually creates two pivotal points around which the swing is then flattened?

3. *Moving forward over a bent forward knee:* This helps flatten the swinging arc because bending the knee makes it possible to continue to move forward and allows for a greater transfer of weight through the action and follow-through phases. It also lowers the center of gravity and keeps it over the base of support, which also improves one's balance. Remember, even a slight wobble can affect direction and accuracy.

The swinging arc can be flattened by using four basic techniques: transferring weight, leading with sequential body parts, moving forward over a bent forward knee, and reaching out during the follow-through.

4. *Reaching out in the desired direction during the follow-through:* Reaching out contributes to the continued flattening of the swinging arc. Concentration on following through in the desired direction is vital. Its contribution to both flattening the arc and the level swing of the impact force is very important. A complete follow-through also plays a major role in ensuring that slowing down does not occur during the force development stage.

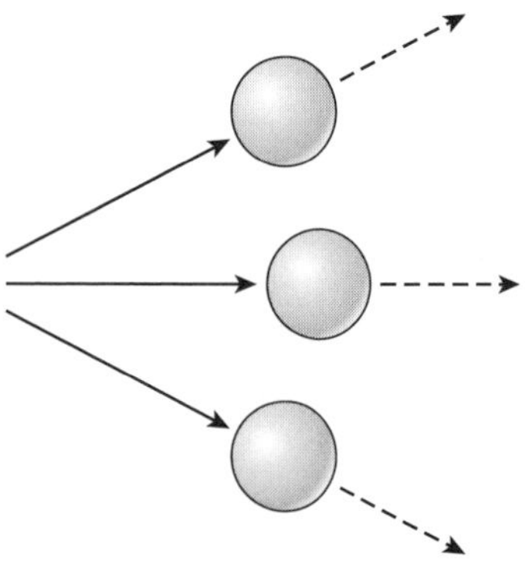

FIGURE 8.14 An object that is hit through the center will travel in the direction of the force impact.

Direction of Impact Force

An object will accelerate in the direction of the force of impact (figure 8.14). If the direction of impact force is a "scoop" or "chop," the resulting path of the object will be considerably different than if the direction of impact is level. If the direction of impact does not pass directly through the center of gravity of the object being struck, a spin will occur, and its effect will contribute to the resultant direction. (See the section on spins in chapter 7; also see figures 7.3 and 7.4.)

Contact Point

If the direction of the impact force passes through the center of gravity of the object being struck, the contact point can become an instructional focal point (figure 8.15).

Here again we are dealing with an equal and opposite reaction. If a person hits under an object, the object will go up, and if an object goes up, the person has hit under it. Being aware of this should help a player to avoid repeating ineffective actions and applications of force. This can then be applied to all directions—down, right, left, and so on.

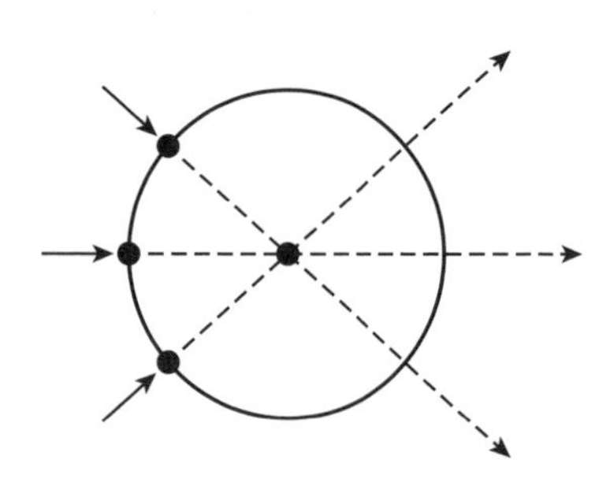

FIGURE 8.15 When the force goes directly through the center of the ball, the contact point determines where the ball will go. Thus, if the ball goes too high, we know that the ball was contacted too far underneath it.

The contact point on the object may be one of the most effective points of concentration for many participants. It gives a relatively clear and specific single point of reference. Other factors, such as kinesthetic awareness (feeling and sensing the position of your body without visual assistance) or the difficult task of evaluating a moving striking implement, may be too complex for many participants.

To modify the contact point, a player has several options:

1. Reposition the object (as in a volleyball or tennis serve; see pages 75-76).
2. Reposition herself in relation to the object (as in golf or tennis; see figure 8.1 on page 75, figure 8.2 on page 76, and figure 8.8 on page 78).
3. Focus and concentrate on the desired point of contact (as in striking by batting, kicking, or racket strokes).
4. Check to see if she is using the techniques that flatten the arc (see pages 77, 79, 80).
5. Change her timing (meet the ball either sooner or later in the swing, as shown in figure 8.16).
6. Check the need for a change in her execution, such as (a) is she swinging level (figure 1.2 on page 7), and (b) is she waiting until the ball toss comes down too low on the tennis serve?

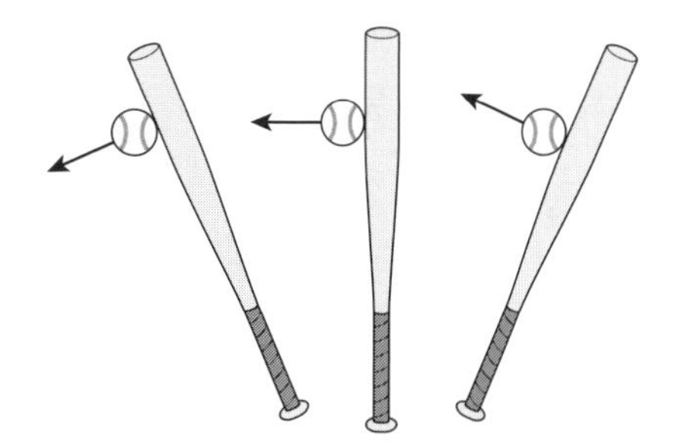

FIGURE 8.16 As the bat swings, the contact point on the ball changes. If the batter wants the ball to go right (e.g., right field), should the batter try to contact the ball early or later in the swing? Would this be different for a right- or left-handed batter?

In deflection sports such as bowling, pool, and billiards, a participant needs to have a thorough knowledge of the results of various contact point impacts and their resulting deflection patterns. This is especially true when one object must be hit in order to hit another object.

Angle of Rebound or Striking Surface

The angle of the rebound or striking surface (the direction in which the surface is facing) affects the rebound direction of an object hitting this surface. This is true whatever the surface is: wall, floor, racket, club, hand, or foot (figure 8.17).

Think about how the angle of the rebound or striking surface relates to various sports—for instance, hitting in billiards, pool, golf, tennis, and badminton; kicking in soccer; wall or ceiling shots in racquetball, handball, or squash; and backboard shots and bounce passes in basketball. Can you see how this principle applies to blocked shots such as a net or half volley in tennis, a spike in volleyball, or a trap in soccer? Can you judge the angle of the ball coming off the ground (bouncing) in baseball, softball, tennis, racquetball, handball, or squash?

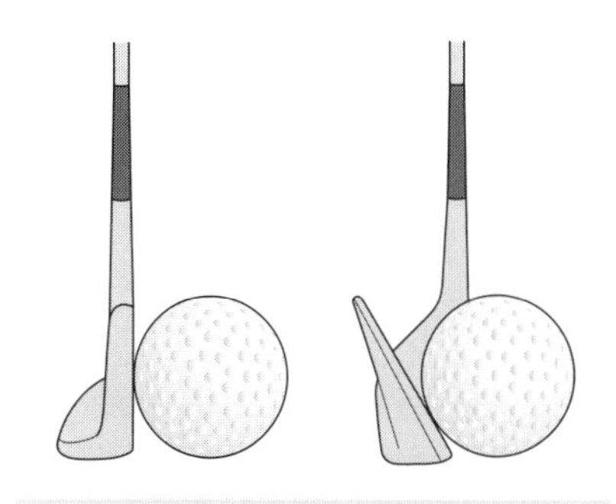

FIGURE 8.17 Can you see why a golfer has a bag full of irons with different angles? Golfers choose their club according to the angle of the club head and where they want the ball to go.

The Action–Reaction Principle

As we have previously discussed, "to every action there is an equal and opposite reaction." If you want to run, walk, or swim forward, then whenever possible all your push or pull should be in a backward direction. Any push or pull not in line with and opposite to the desired reaction can lead to less desirable results. Can you see how toeing out or in could

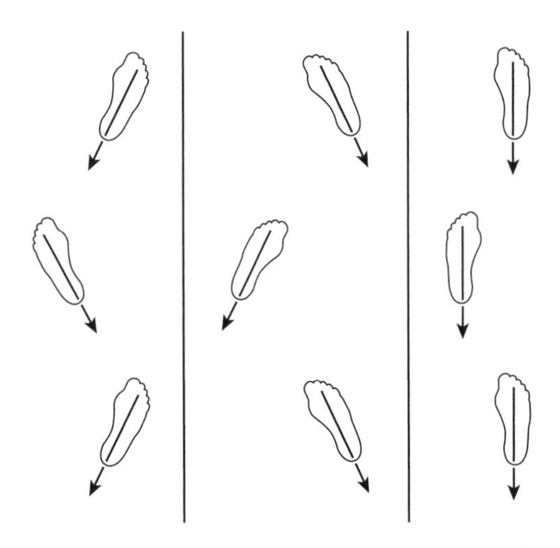

FIGURE 8.18 Which of these is the most effort efficient and will send you in the direction you want to go?

lead to wasted effort (figure 8.18)? Could you analyze the effects of various swimming strokes in relation to this principle?

Can you see how the action–reaction principle also relates to angles of rebound, contact points, direction of impact force, flattening of the swinging arc, and timing of the release or striking point?

Spins

For information on spins, review chapter 6, pages 58-61.

Resultant Force

Forces are frequently found working in combination with other forces. Only occasionally will a single force be in operation in a given situation. When forces work together, we have what is known as a resultant force—and in turn we have a resultant effect.

An example might be a racket contacting a ball. In this example, the result may be affected by

1. the direction of the impact force of the striking implement,
2. the point of contact,
3. the angle of the rebound surface, and
4. any spin on the ball at the time of impact.

Figuring out resultant forces and their effects can be fun and challenging. The easiest method may be to begin by determining the additional effect that each force would contribute and how this would change the usual result. We have already done this in part by observing the effects of spins on the normal rebound pattern.

Understanding and predicting possible effects of various factors on direction may help you alter your techniques. This will become more essential as you seek to become highly skilled or to coach the skillful player.

Four factors that can negatively affect our ability to execute directional control are (1) the inability to stabilize or set body parts that create a foundation against which moving parts can push or pull; (2) a lack of good balance; (3) excessive tension; and (4) extraneous movements.

The Participant's Body

Four factors that can negatively affect our ability to execute directional control are (1) the inability to stabilize or set body parts that create a foundation against which moving parts can push or pull; (2) a lack of good balance (the foundation from which we direct force); (3) excessive tension, which can inhibit our ability to execute appropriate techniques; and (4) extraneous (unnecessary) movements, which may contaminate directional application of force. These were discussed in chapters 2 and 4.

Observations

• Observe the effect of gravity. Push or toss a ball into the air. Watch it carefully. Can you see how it slows down as it rises and pauses slightly at the peak of the toss? As the ball moves downward, the speed with which it is traveling is increased by the continuous pull of gravity. At what point would a beginner find the ball easiest to contact in relation to the speed at which it is moving? Can you relate this to a tennis serve?

• *Go to a bowling lane and observe the path of the ball of a proficient bowler.* Note how the bowler flattens the swinging arc by getting down over a bent knee, transferring the weight forward, and reaching out on the follow-through. Try it.

Attempt to find a bowler who drops or bounces the ball onto the alley. Can you diagnose why this occurs? Can you see why a smooth bowling release would provide greater consistency and thus greater accuracy? (See figure 8.11 on page 79.)

Observe a hook ball roll. Can you determine how the hand position puts spin on the ball even without any turning of the hand? (See page 59 and figure 6.4 on page 59.) Can you see why trying to put spin on a hook delivery by twisting the hand or lifting the wrist might prove ineffective for the beginner or inexperienced bowler?

Activities

• Loft several paper wad balls into a basket several feet away. Are you (almost without being aware of it) developing a technique based on the pattern of the pull of gravity? Note your follow-through. Are you swinging your hand and arm directly out toward the basket? This ensures a good direction of flight. Observe the follow-through of other people. Observe the follow-through in kicking and striking activities. Can you see why the follow-through in shooting activities, such as archery or riflery, is a "hold" or "freeze" in the final position and is absolutely vital to accuracy?

• Take a light weight ball—such as a beach ball, Wiffle ball, or table tennis ball—and attempt to spin it in various directions as you throw it forward. Can you observe the various flight patterns in relation to the different spins? Can you explain why each of these occurs (see page 66 and figure 7.3 on page 66)?

• Practice leading with each sequential body part to flatten the swinging arc. Can you see why this affects the direction of a projectile?

• Place a four-foot length of plain shelf paper on a table. Facing the table, grip a marking pen so that you can draw the path of your hand as it passes along the table. Attempt to flatten your moving hand and arm as you do the following:

1. Transfer your weight from your back foot to your front foot as you move your arm forward.
2. Repeat without any weight transfer.
3. Lead with each sequential body part.

4. Repeat, keeping the joints straight; don't lead.
5. Have your forward knee bent as you move your weight forward.
6. Repeat, keeping your knee straight as you move forward.
7. Try reaching out on the follow-through.
8. Repeat, but do not reach out on the follow-through.

Can you see the effect on the flattening of the arc as you attempt each of these? If you have difficulty differentiating the various trials, you might want to use different colored marking pens.

• Tee a ball (see figure 8.19). You might also choose to hang a tarp or heavy cloth to hit into. This could be marked with target areas.

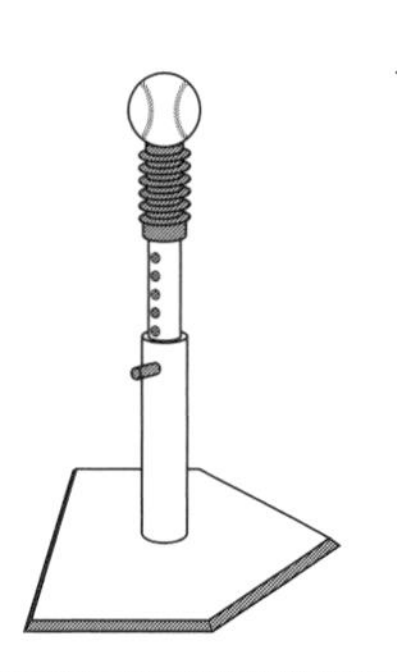

FIGURE 8.19 What is the difference between swinging level (figure 1.2 on page 7) and flattening the swinging arc (figure 8.13 on page 79)? Could you practice both using the Batting Tee Challenge in chapter 10?

Using various directions of impact, including chopping and scooping actions, hit the ball off the tee. Note the effect. Now practice hitting with a level (horizontal to the ground) swing. If the ball rises more than you want it to, what should you check about your swing and the contact point? What if the ball goes down? What if it goes left or right? Can you see how this latter problem involves the flattening of the swinging arc rather than the levelness of the swing to the ground? Can you separate these two problems and see how they differ?

Hit the teed ball at several different contact points. Can you see how this affects the direction the ball takes? Can you reverse your analysis and go from result to cause (for instance, the ball went up, thus it must have been hit from below)?

Hit the ball off center. The result should be a spin. Can you determine the type of spin imparted and the directional effect of each type of spin?

• Try to find a means by which you can alter the angle of a rebound surface (see Rebound Board in Appendix B on page 133). Experiment and observe the results when the angle of the rebound surface is changed. Can you predict the results? When would this information be important?

• Take a pole that has sufficient length yet is light enough that you can hold it up high without excessive fatigue. An old bamboo fishing pole works well. Tie a string or fishing line to the pole, and attach a paper clip to the free end of the string. Now place the plastic part of a badminton shuttlecock in the clip as shown (figure 8.20). This can also be done with a tennis ball using Velcro.

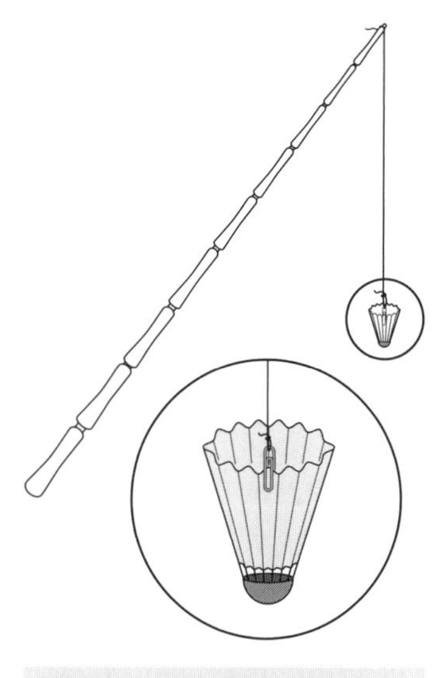

FIGURE 8.20 The use of a paper clip to attach the badminton bird to the string allows the bird to be released on contact during the swing.

Have the bird held at three different points: out in front of you approximately 1 1/2 to 2 feet (46 to 61 cm), directly over your head, and slightly behind you (figure 8.21).

The bird should be held high enough to force you to take a *full* reach as you hit it from each position. Note the direction of flight, the contact point on the bird (under, behind, above), and the angle of the striking surface. To identify the angle of the striking surface, you may have to hold your racket up to the bird without striking it. Can you draw any conclusions from this experiment? *Note:* If you must work alone, tie the line or string overhead and then place yourself in various positions under the bird.

Try to relate this experiment to (1) a tennis serve, (2) an overhead volleyball serve, (3) an underhand volleyball serve, (4) a spike in volleyball, (5) a golf drive, (6) a smash in tennis or badminton, (7) a *low* short serve in badminton, and (8) a drop shot in badminton or tennis.

If a golfer wanted to hit his golf drive lower, which direction would he need to move in order to accomplish this?

If a left-handed batter wanted to hit a ball to right field in baseball or softball, at what point in the ball's travel to home plate should the batter attempt to hit the ball (early or late)? Would this change if the batter was right handed?

FIGURE 8.21 Where does the bird need to be contacted to execute a smash? If the bird goes straight ahead when you wanted to execute a smash, what does this tell you about your position in relation to the bird at the point of contact?

Can you complete the movement patterns in the following figures? Before attempting figures 8.22 through 8.31, you might want to review the information about spins on pages 58 to 61.

Remembering that a spin is named by what is happening to the front of the ball, try to identify and predict movement patterns in figure 8.28 through 8.31.

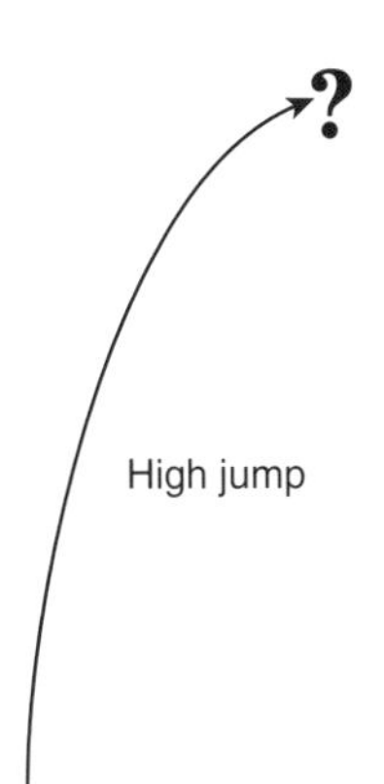

FIGURE 8.22 Can you draw in the rest of the path of the center of gravity within the body?

FIGURE 8.23 Complete the path of the flight of the badminton bird. Don't forget about air resistance.

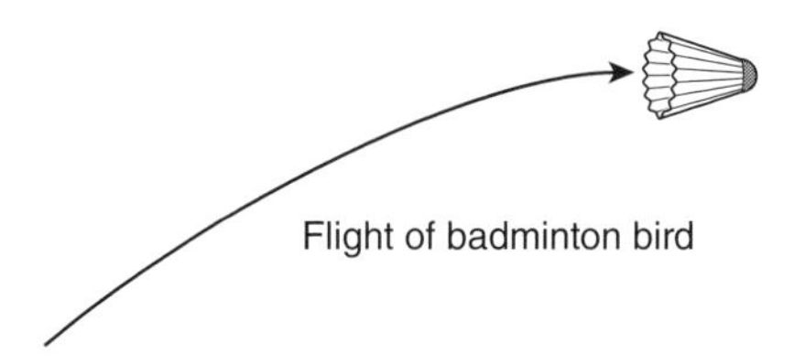

FIGURE 8.24 Draw in the path of the ball as it comes up from the floor or ground. Which way will the spin push against the ground? What will that cause the ball to do? Be sure to deal with height, speed, and distance.

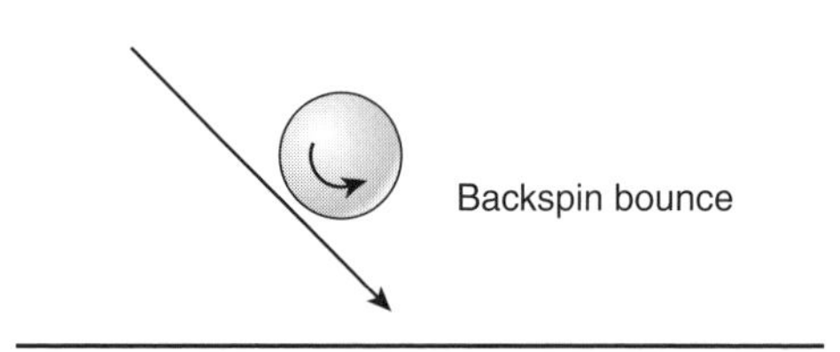

FIGURE 8.25 Draw in the path of a left-handed hook ball roll. Could you explain why this happens?

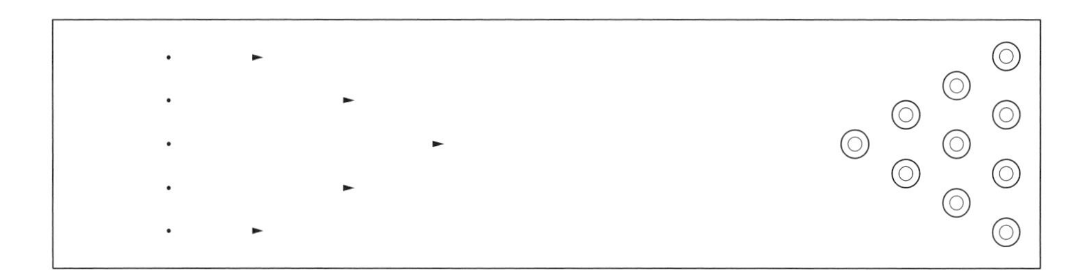

FIGURE 8.26 What kind of spin would you want on this ball to make it bite and not roll past the cup?

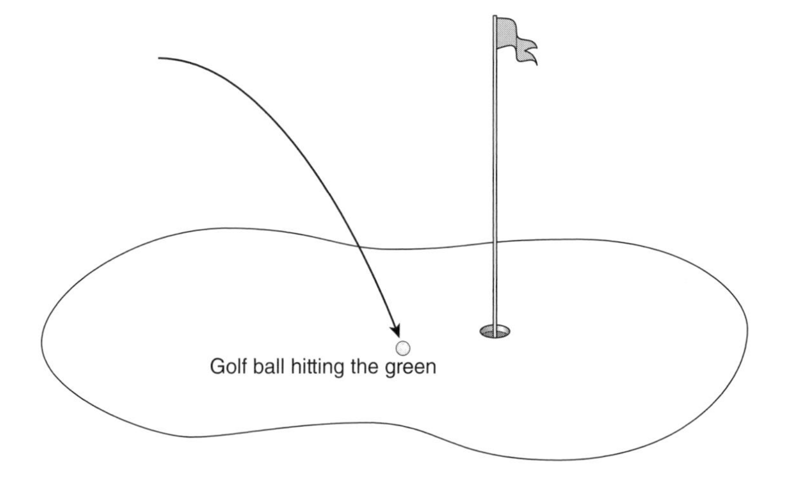

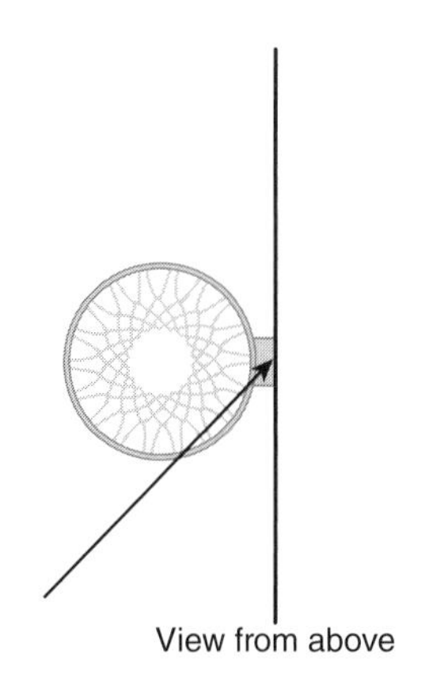

FIGURE 8.27 Draw in the rebound path for this basketball backboard shot when there is no spin on the ball. Can you see the similarity to the bounce from the floor with no spin on the ball?

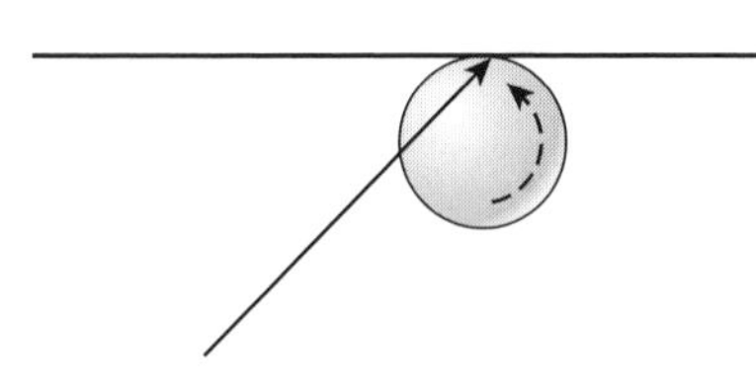

FIGURE 8.28 Think of this as a ball hitting a basketball backboard. Write in the spin name. Above the ball, draw in an arrow showing which direction the push against the backboard would be. Then indicate the path that the ball would take as it leaves the backboard.

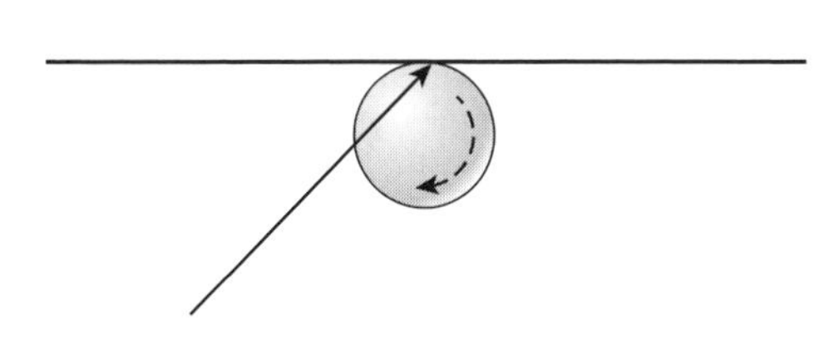

FIGURE 8.29 Think of this as a ball hitting a basketball backboard. Write in the spin name. Above the ball, draw in an arrow showing which direction the push against the backboard would be. Then indicate the path that the ball would take as it leaves the backboard.

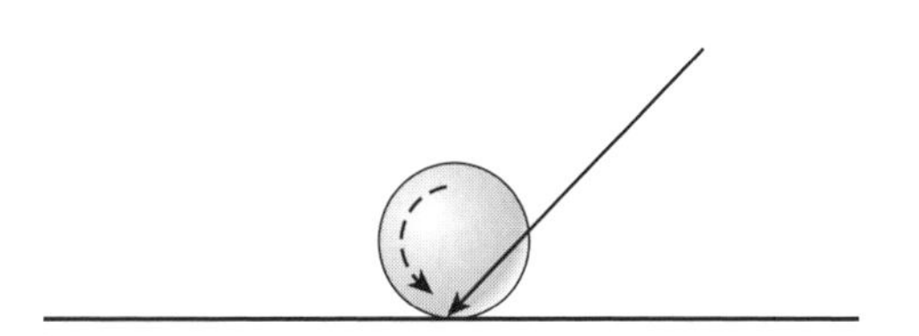

FIGURE 8.30 Think of this as a basketball hitting the floor. Write in the spin name. Above the ball, draw in an arrow showing which direction the push against the floor would be. Then indicate the path that the ball would take as it leaves the floor.

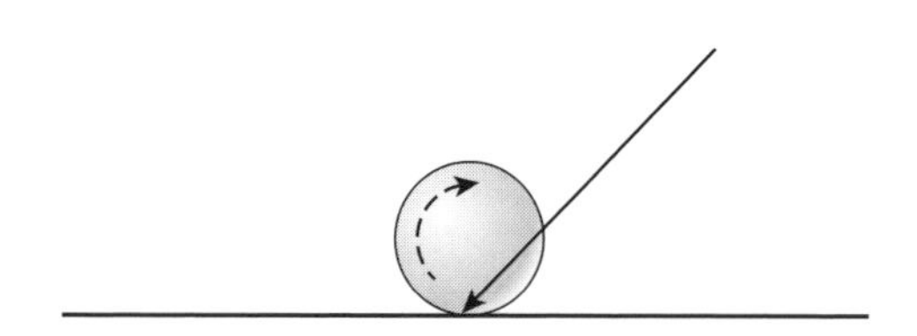

FIGURE 8.31 Think of this as a basketball hitting the floor. Write in the spin name. Above the ball, draw in an arrow showing which direction the push against the floor would be. Then indicate the path that the ball would take as it leaves the floor.

True or False Review Questions

Information about each question is found on the page number provided within the parentheses following the question.

REMEMBER THAT A PARTIALLY FALSE QUESTION IS CONSIDERED FALSE. Answers to the questions and additional information are found in appendix C.

1. It is possible to analyze results by using the "curious child" method of "why, why, why" to discover the cause of errors (pages 4-5, 75).
2. We need to pay attention to all forms of air resistance. Three forms of air resistance are wind, drafts, and spins. The latter one is the only one we can use to our advantage. The other two create only disadvantages (page 77).
3. Flattening the swinging arc refers to swinging level *in the same plane* (pages 7 [numbers 4 and 5] and 77-80).
4. Flattening the swinging arc affects both accuracy and force development (page 32; also see figures 8.9 and 8.10 on page 78).
5. Flattening the swinging arc is good for more advanced players, but it has no real benefit for the novice or beginner (pages 32, 78).
6. The four basic techniques used to flatten the swinging arc are transferring weight in the preparatory, action, and follow-through phases; leading with sequential body parts; moving forward over a bent forward knee; and reaching out in the desired direction during the follow-through (pages 79-80).
7. Moving forward over a bent forward knee helps flatten the swinging arc because bending the knee makes it possible to continue to move forward and allows for a greater transfer of weight through the action and follow-through phases. It also lowers the center of gravity and keeps it over the base of support, which improves one's balance (page 80).
8. A good follow-through ensures that a slowing down and an absorption of force will occur during the final part of the action phase (pages 5, 37, 80).
9. The action–reaction principle relates to the efficient use of force in the desired direction. It also relates directly to angles of rebound, contact points, and direction of impact force. This principle relates indirectly to the flattening of the swinging arc and to the timing of the release or striking point (pages 81-82 and figure 8.18 on page 82).

10. In highly skilled or advanced play, most forces are actually resultant (combination) forces that have resultant effects. Predicting the possible overall effect of all the force and directional contributors requires an attempt to determine the additional effect that each contributing factor would have on the result (page 82).

11. Four factors that can negatively affect our ability to execute directional control are (1) the inability to stabilize or set body parts that create a foundation against which moving parts can push or pull; (2) a lack of good balance (the foundation from which we direct force); (3) excessive tension, which can inhibit our ability to execute appropriate techniques; and (4) extraneous (unnecessary) movements, which may contaminate directional application of force (page 82).

12. The direction of spin to be put on a layup shot in basketball should be the same regardless of whether the approach is from the left or right side of the basket (spins on pages 58-61, resultant force on pages 58-59, 82).

PART IV

Improving Analytical Skills

CHAPTER 9

Visual Evaluation

Good visual evaluation can be extremely helpful. It can assist you in predicting what will happen and allow you to be ready for it. I have found that good visual evaluation is second only to good balance in contributing to success in sports and play.

Beginners must learn to "keep their eye on the ball." Intermediate players need to become proficient at evaluating the path of the ball, assessing its travel speed, and making visual estimations of distance. They must also learn to judge and predict rebounds (including bounces) and begin to read the movements of an opponent. Advanced players become sensitive to early indicators that will help them determine and adjust to unusual conditions. These players can make effective predictions that will allow them to be where the action is, and at a quick glance, they can read an entire situation to make important split-second decisions. These skills are developmental and require experience. But being aware of the basic principles discussed in chapters 6, 7, and 8—and using carefully selected and related activities to gain experience—can help to reduce the time needed to acquire and improve these skills.

Evaluation Skills and How to Improve Them

Here are some specific visual skills that can be developed or improved:

1. Visual focus and concentration
2. Visual tracking—the ability to follow the movement of an object or person over time through space
3. Reading and predicting movement of an object, an opponent, or a situation
4. Visual figure and ground discrimination—the ability to separate the relevant stimuli (figure) from the competing irrelevant stimuli (ground) in order to focus on the relevant factors
5. Relating and evaluating multiple visual stimuli simultaneously
6. Judging speed and distance

Games such as Chaotic Team Juggle (found in chapter 10, pages 103-104) can be used to practice all six of these visual skills. Other games that are useful for practicing these skills include Blanketball, Frantic Ball, Living Basketball, Partner Scoop Play, Pass-a-Puck, Racket and Balloon, Rag or Rug Hockey, and Rebound Board.

Visual evaluation can be improved with practice, especially if that practice is planned to be specifically enriching and instructive. For example, individuals must be able to concentrate their focus on an object to track it visually as it travels through space. Without the former skill, the latter becomes much more difficult and may remain weak. By concentrating their focus first on a still object, then on one that is moving slowly and in a predetermined pattern, the individuals can progressively improve the set of skills required to track an object at various speeds and variable patterns (refer to the discussion of progressions in chapter 1, page 6; also see figure 1.1 on page 6).

Following are some additional suggestions, but you need to be analytical and inventive yourself. You may discover methods, games, and techniques that are even more helpful.

- Tee an object on a holding device of some sort and let the player hit it. (Information on how to make a batting tee is provided on pages 128-129 in appendix B.)
- Tether or tie an object or ball so it can swing free. First allow a player to hit the ball while it is hanging still. As success increases, add a challenge by swinging the ball ever so slightly in line with the swinging pattern of the hitter. Then allow the player to see how many times in a row she can hit the swinging object. You can also tether an object with a clip device (or the creative use of Velcro) so that a solid contact will release the object and hitters can see the results of their concentration and tracking. If contacting the moving object is difficult for the player, have the player work on improving visual concentration and focus first. One way to do this is to mark the tethered ball with letters, colors, or words. Then, giving a very slow spin to the swinging object, ask the player to attempt to identify or read what is on the ball.

Remember, skills are developmental, and our task is to find what aspect or aspects need help and how we can creatively and effectively extend the needed help. A player's optimal ability may never be reached if the foundation skills are left incompletely developed or are neglected because the player can hit or catch a ball and is having some success. We need to find enjoyable activities to give each participant challenging experiences that allow for continuous growth and development (see chapter 10).

Rebound devices are available that allow a player to practice catching by himself when another player is not available. The side of a building or a wall has served this purpose for almost as long as there have been walls and balls.

Visual evaluation can be improved with practice, especially if that practice is planned to be specifically enriching and instructive.

You may want to change the size or color of an object to assist the individual who is having difficulty. Perhaps the object is traveling too fast. Playing with a beach ball or balloon may help. Juggling certainly encourages continuous improvement of visual concentration and tracking, and it can be a challenge to the more highly skilled player.

Another important skill related to visual evaluation is the ability to read one's environment and others' movement patterns to gain early cues—and thus to be able to predict and respond more successfully.

Another important skill related to visual evaluation is the ability to read one's environment and others' movement patterns to gain early cues—and thus to be able to predict and respond more successfully. This could involve predicting where an object will land, how fast it is traveling, the angle at which it will rebound, where another player is located, what that player's next move might be, whether an opponent has put a spin on the ball, and how the flight and rebound will be affected by various conditions.

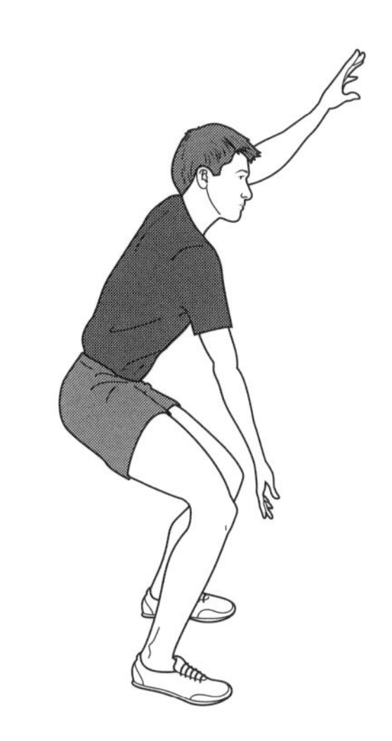

FIGURE 9.1 Learning to continuously visually evaluate your environment is important in play and sports, and also in life.

Activities

- Begin with slow mirroring or shadowing. This is a process where one person or a group attempts to follow and duplicate all the movements of another (figure 9.1).
- To increase the challenge, simulate a basketball situation that involves guarding a player. The situation should include stops, starts, and attempts to lose the guarding player.

When seeing a skillful player making a well-coordinated response, many people credit it to "natural ability," unaware of the developmental practice that helped the player build this ability to respond effectively. Some people also see these skills as useful only during play. But developing the ability to evaluate visually, adapt, and respond can help people in various other situations. For example, a person who can make good visual evaluations might be able to avoid a collision in a possible automobile accident or may simply be better able to park a car in a small space.

True or False Review Questions

Information about each question is found on the page number provided within the parentheses following the question.

REMEMBER THAT A PARTIALLY FALSE QUESTION IS CONSIDERED FALSE.

Answers to the questions and additional information are found in appendix C.

1. Good visual evaluation is very important. Beginners are encouraged to keep their eye on the ball. Intermediate players can begin to evaluate the path, speed, and distance of the object. Advanced players can become sensitive to early indicators that will allow them to read their environment and make earlier predictions, split-second decisions, and better adjustments to unusual conditions (pages 93, 95).

2. An understanding of the mechanical principles of movement will not reduce the time and practice needed to improve skills, but it will allow players and coaches to do a better job of selecting the proper techniques (page 93).
3. Visual tracking refers to following the movement of an object or person over time through space (page 93).
4. Visual figure and ground discrimination is a learned skill that involves having the ability to ignore the irrelevant factors in a visual field in order to focus more directly on the relevant factors (page 93).
5. If a player is successful but is weak in one of the foundation skills (such as balance or visual figure and ground discrimination), this weak skill will probably develop or improve through regular participation and can be ignored as a problem for now (page 94).

CHAPTER 10

Activities to Habituate the Mechanical Principles of Movement

Look beyond the sport-specific drills and work on the underlying issues that are based in the mechanical principles of movement.

Play May Be the Key to Open Many Doors

To players: Having fun with the activities in this chapter will also help you become intuitive about the application of the seven mechanical principles discussed in this book.

To elementary, middle school, and high school physical education teachers: The games in this chapter are ideal activities to insert into lesson plans and can be used as warm-ups, cool-downs, or ways to practice the mechanical principles being taught in that lesson. In high school, a student unable to actively participate could study a chapter from this book and answer the questions at the end of the chapter. The answers are found in appendix C.

To college and university physical education teachers: The activities in this chapter can provide a good supplement for your classes. They could also be done as an independent study or for extra credit.

To coaches: The games in this chapter could be used for some end-of-practice fun while serving as cool-down activities. They could also be used to promote additional cardiovascular development or to practice mechanical principles that will improve players' playing abilities. You may even want to encourage players to play these games at home.

To parents: You can use these activities to help your children find increased success in sport and play. These activities can also provide a fun family time together. Your children may never forget these times and may even play the games with their children and their grandchildren.

To high school and middle school science teachers: These activities could be an interesting way to involve high school students in the science of biomechanics.

This chapter focuses on seven important mechanical components and provides games and activities that enhance their application (see Game Matrix on page 100):

- Balance
- The effects of specific impact (contact) points on a round object
- The effects of various angles of a rebound or impact surface on a rebounding object
- Flattening (lengthening or extending) the path of a swinging body part or striking implement for increased accuracy and force
- Follow-through
- Absorbing force
- Reading the parabolic path. If you are unfamiliar with this component, refer to the definition (and figure 7.1) on page 65 in chapter 7.

Why do I encourage you to use the activities in this chapter rather than just play any games or simply practice specific sport skills? The games and activities in this chapter have been carefully selected because they draw responses from the players that help them discover (or intuit) the mechanical principles discussed in this book. Through the simple repetitive play in these activities, these principles become habituated by the players to the point that they can be transferred and applied to more complex movement challenges.

This is somewhat like learning to walk. Before we can stand to take those first steps, we need to spend hours trying to roll over, sit up, and crawl. Only after we have gained all the basics that need to be automatically and quickly available to us will we risk—usually with the help of something or someone to cling to—attempting the newfound and more complex skill of walking. No one that we know of has taken those first precarious steps successfully without first conquering the prerequisites. The seven mechanical principles are the prerequisites to moving well and successfully.

Through the simple repetitive play in these activities, the mechanical principles become habituated by the players to the point that they can be transferred and applied to more complex movement challenges.

Balance is the keystone of all movement and of the ability to hold or control one's position. Therefore, balance should be emphasized when helping a child or youth prepare to become "natural" or coordinated in his movements. We often assume that an individual will develop good balance. Many do not. We can help players at any age improve their balance by getting them involved in games such as Skates, Freeze, Frantic Ball, Super Sox, and Rag or Rug Hockey. (Note the story on pages 20-21 in chapter 2.)

Balance, follow-through, flattening of the swinging or striking arc, and absorbing force can be improved by bending the knees. Holding a bent-knee position requires both strength and endurance of the lower limbs. The strength and endurance needed can be gained without verbal instruction through games such as Frantic Ball, Skates, Super Sox, Freeze, Pass-a-Puck, and Rag or Rug Hockey. These games also encourage players to bend their knees in other activities, which will increase their success.

Games such as Living Basketball, Blanketball, Rebound Board, and Partner Scoop Play give players lots of practice in reading the parabolic path of an object. Learning to do this will help a player have more success in catching, throwing, shooting baskets, aiming, or receiving a projectile. These same experiences also make a contribution to a player's ability to effectively flatten the swinging arc, follow through, and absorb force.

To provide further practice in absorbing force, challenge players to run, hop, or jump softly or quietly so you cannot hear them; adjust to the oncoming force in Super Sox; and control their momentum in the quick stops in Freeze.

Players can improve their batting or golf swing and learn to combine all the elements that flatten the swinging arc (sequential swing, transfer of weight, bending the forward knee, and follow-through) in the Batting (or Golf) Tee Challenge. This same practice and equipment can also help players increase their awareness of the effects of various impact or contact points. All of these may then become available choices to apply and adapt to the various movement patterns in sports that involve the flattening of the swinging arc (hitting, throwing, pitching, and specific sport skills in golf, tennis, handball, racquetball, volleyball, bowling, softball, baseball, football, ice and field hockey, and so on).

The flinging action in Pass-a-Puck and Rag or Rug Hockey also enables players to practice flattening the swinging arc along the floor or ground. This can then be applied to related sport skills such as ice and field hockey, soccer ground passes, and the delivery in bowling.

In the game of Frantic Ball, the challenge to keep the balls on the floor encourages players to practice the flattening (lengthening or extending) or smoothing out of the path of the swinging foot, the contact or impact point on the ball, and the angle of the surface of the striking foot. In this game, players will also practice balance over a small base of support (one foot—as the other foot is swung).

Awareness of how specific contact points and various angles of an impact or impacted surface affect the rebound flight of a projectile is important to all related skills that involve impact force. Activities such as Racket and Balloon, Frantic Ball, and Rebound Board allow a player to gain experience that can contribute to this awareness.

Balance is the keystone of all movement and of the ability to hold or control one's position. Therefore, balance should be emphasized when helping a child or youth prepare to become "natural" or coordinated in his movements. We often assume that an individual will develop good balance. Many do not.

You can make some of the equipment discussed in this chapter. See appendix B for more information. Encouraging players to make their own equipment (e.g., batting tees, rebound boards, newspaper-packed balls, racket and balloon, skates, scoops, or sticks and button bags or beanbags for Pass-a-Puck and Rag or Rug Hockey) at home may increase the amount of practice and may thus improve the skills and awareness being developed.

GAME MATRIX

GAMES (m) = modifications available	Page number	Absorbing force	Angles of rebound surface	Balance	Contact points on a round object	Flattening the swinging arc	Follow-through	Reading the parabolic path
Batting Tee Challenge	101			●	●	●	●	
Blanketball (m)	102	●	●				●	●
Chaotic Team Juggle (m)	103	●					●	●
Frantic Ball (m)	105	●		●	●	●	●	
Freeze (m)	107	●		●				
Golf Tee Challenge	108		●	●	●	●	●	
Living Basketball	109	●					●	●
Partner Scoop Play	110	●				●	●	●
Pass-a-Puck (m)	111			●		●	●	
Racket and Balloon	113		●		●	●	●	
Rag or Rug Hockey (m)	114			●		●	●	
Rebound Board	116		●		●	●	●	●
Skates (m)	117			●				
Super Sox (m)	119	●		●				

Batting Tee Challenge

The Batting Tee Challenge is great for improving batting, but it can also improve a player's skills in hockey, soccer, golf, tennis, or any other sport that involves hitting or kicking an object as it habituates the flattening of the swinging arc. Appendix A (Putting It All Together: Some Suggestions on Striking) contains useful information that is relevant to this activity.

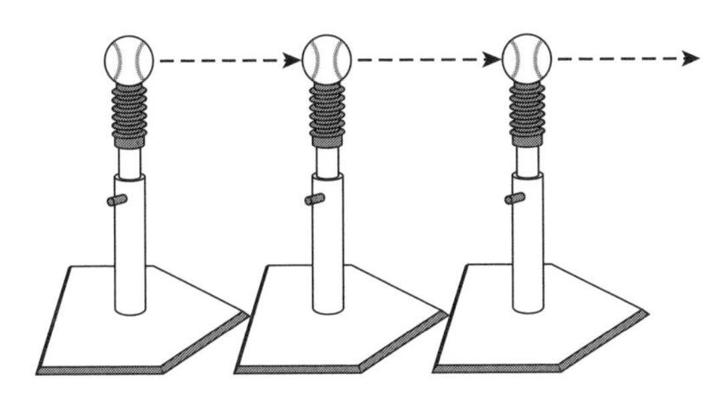

This challenge will encourage the batter to both swing level and flatten the swinging arc.

Equipment Needed

- Two or three adjustable or homemade batting tees (see appendix B for instructions on how to make batting tees)
- A bat made of tightly rolled newspaper (see appendix B for instructions)
- Several well-packed newspaper balls. Having several balls will allow players to have several turns so they can evaluate problems and try again. Making these out of well-packed newspaper then taping them so they'll hold together will also allow for absorption of force, reducing the space needed (see appendix B for instructions on how to make these).
- A sheet to be hung and hit into (optional)

Mechanical Principles Being Practiced

- Moving balance (as the player learns to adapt his center of gravity and base of support in relation to the demands of the task)
- Increasing sensitivity to the effects of contact or impact points
- Flattening of the swinging arc by bending the forward knee, transferring weight, sequentially timing the use of each involved body segment, and using a good follow-through to hit the last ball. This supports the development of increased force and accuracy.
- Follow-through (especially as the number of tees increases from one to two to three)

Instructions

If a batter can hit one ball off one tee, challenge him to see if he can hit two balls off two tees with one swing. Next, have the batter try three tees. Initially, the tees can be lined up close to each other. As the batter becomes more skillful, the tees can be moved farther apart. If practicing at home, a batter can hit into a sheet or blanket hung over a rope line in the backyard, basement, or garage.

This challenge encourages players to keep their eyes on the ball. It also encourages them to use a sequential swing involving all appropriate body parts, a bent forward knee on the forward swing, a good weight transfer, and a good follow-through. These actions assist the hitters in flattening the swinging arc, which helps them develop increased force and accuracy.

Comments and Suggestions

- Flattening of the swinging arc can lead to greater force and accuracy in many sport skills. Participants can transfer the components involved in the flattening of the arc to tennis, golf, bowling, soccer, hockey, and so on.
- By practicing with multiple batting tees, the hitter is habituating the body to automatically perform the mechanical principles that are common to all sport skills that involve flattening of the swinging arc.
- Having more than one teed ball to hit also encourages focus and concentration.

Blanketball

Blanketball is a group challenge, and the players participate as a team. In this game, the players will focus on how to absorb force and how to read the parabolic path. Players will become increasingly aware of the effects of various angles of the force-producing surface as they read and predict the flight of the object.

Equipment Needed

- A sheet, bedspread, or other light material for each group of approximately 8 to 12 players (depending on the size of the material)
- A soft object such as a paper-packed ball made by loosely wadding newspaper into a ball and securing it with rubber bands or tape—one for each group
- If using the Towel or Pillowcase Toss modification, a paper-packed ball or other soft object and a towel or pillowcase for each pair of players

Mechanical Principles Being Practiced

- Learning how the angle of the rebound surface affects flight
- Reading the parabolic path to send and receive an object
- Absorbing force (give)
- Follow-through

Instructions

Each group of 8 to 12 players arranges itself around the material and holds the edge of the fabric. A soft object is placed in the center of the material. The group is challenged to do each of the following, one at a time. When one challenge is conquered, instruct the group to move on to the next one.

1. Lift the object just 6 inches (15 cm) off the material, and when it comes down, stop it without letting it bounce (absorbing force).
2. Alternate lifting and stopping the object—first 6 inches (15 cm), then 6 feet (1.8 m) (absorbing force).
3. Lift the object as close to the ceiling as possible without touching the ceiling (force control).
4. Lift the object on an angle (angle of rebound surface) and then move the sheet under it as it comes down (reading the parabolic path).

5. Toss one object between your group and another group (angle of rebound surface as the group tosses to the other group, reading the parabolic path in order to receive the object, and absorbing force as the received object comes down onto the material).
6. Toss and exchange two objects between your group and another group (angle of rebound surface, reading the parabolic path, and absorbing force).
7. Send an object through a basketball hoop (angle of rebound surface and determining the needed parabolic path to drop the object through the basket).

A group does not need to accomplish all seven of these challenges during any one session.

Modifications

- Towel or Pillowcase Toss: Two players (working as partners) can use a towel or pillowcase. This is played the same as Blanketball, and the same mechanical principles are practiced. The same progression can be used.
- In teams of 6 to 8, play the game like volleyball, using a piece of yarn as the net.

Comments and Suggestions

- The leader can let individual groups progress at their own pace through the challenges. Be sure that new challenges are available as each group seems ready.
- This is a good game for hot days—the moving material fans the participants.
- This activity enables players to practice all six visual evaluation skills covered in chapter 9. These skills help people read their environment in preparation to solve a problem.

Chaotic Team Juggle

The objective in Chaotic Team Juggle is to see how many objects the team can successfully keep in motion. This is another activity that enables players to practice all six visual evaluation skills covered in chapter 9. Therefore, the players will be learning to read their environment in preparation to solve a problem.

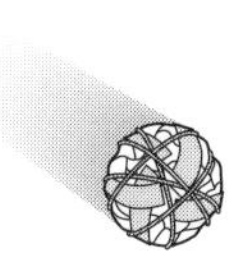
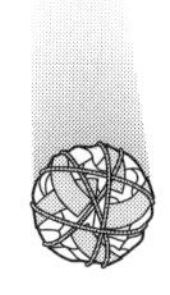
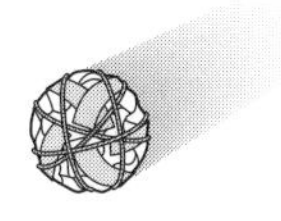

Players use newspaper balls for the Chaotic Team Juggle activity.

Equipment Needed

- 8 to 15 soft newspaper balls—2 for each three players (see appendix B for information on how to make newspaper balls)
- 1 to 3 large trash bags to end activity

Mechanical Principles Being Practiced

- Reading the parabolic path to send and receive an object
- Absorbing force (give)
- Follow-through

Instructions

Before passing out several soft newspaper balls (about one for every three players), explain that each participant with an object may throw it to anyone—*but must first get the person's permission by getting his attention.* Permission is granted only if a thrower can make and hold eye contact with a potential receiver. The group's challenge is to see how many balls it can keep in the air at once. If a ball hits the ground, a player should pick it up and start it again. Let this action go on for a while. It may become chaotic. You may even have to stop this activity. One way to end the game is to hold open (or have one or two players hold open) a large trash bag and encourage players to shoot a basket. This is now referred to as Living Basketball, which is a game found later in this chapter. You may have to remind players about getting "permission" before shooting a basket.

Modifications

- Increase or decrease the number of objects.
- Vary the size, weight, and shape of the objects.
- If you are in a situation where noise will be disturbing, add a rule that players must get each other's attention without any sounds being used. This is called Silent Chaotic Team Juggle.

Comments and Suggestions

- Players may be in a circle or a random formation.
- Newspaper balls, paper wad balls, sock balls, or pillows are good to use because they are soft, safe, and not apt to cause damage if missed. This equipment is readily available and can be repaired and replaced with ease, making it possible to play the game almost any place at any time.
- Players need to concentrate more to adapt to the inconsistent flight of a ball that is not perfectly round. This adaptability is perhaps an essential ingredient in becoming an exceptional athlete.
- If Chaotic Team Juggle becomes too stressful for a player, this player is able to buy out by simply not looking at anyone until he is ready to buy back in.
- To get the most benefit from this activity, have the players repeat it several times over a long time period. Over this time, you might consider emphasizing the following:
  - Follow through directly at your target (receiver's chest).
  - Help your receiver make a successful catch.
  - Be sure you have your receiver's attention *before* throwing.
  - "Absorb" the ball (catch and bring it to your body) so it won't bounce away.
- The use of opposition of arms and legs (see figure 4.17 on page 40) is not stressed because a forceful throw is not needed. When accuracy is the primary objective, using the same side (arm and leg) is more apt to maintain a straight line of force.

Frantic Ball

This activity practices "reading one's environment" in preparation to solve a problem. It is also good for improving balance—the better the balance the better the player. Frantic Ball promotes soccer skills without using drills.

Equipment Needed

- Tennis balls (these can be old and do not need to bounce)—two for every three players

Mechanical Principles Being Practiced

- Balance (balancing on one foot—a small base of support—while swinging the other foot to hit the ball)
- Flattening the swinging arc (trying to keep the ball on the ground will tend to help flatten the swinging arc)
- Contact points (using experience to determine how to avoid lifting the ball by flattening the swinging arc and hitting directly behind the ball)
- Follow-through (using experience to determine how to avoid lifting the ball by having a good follow-through)
- Absorbing force (to reduce the force of an oncoming ball)

Instructions

The group forms a circle. The object of this activity is to keep as many tennis balls rolling on the floor or ground as possible, using only the feet. To practice, have the group deal with just one ball until they seem to be in control of that ball. Ask the group if they think they can handle another ball. Then *gradually* add other balls as the players think they are ready. This allows the group to get the idea of what is to be done and helps prevent chaos from developing. Once the group has experienced this practice, clarify the scoring system. If a ball stops or lifts and is seen by an official, a penalty point is given to the group (the official indicates this by using a little frantic noise). A stopped ball should be put back into motion and is a continuing part of the game. The group tries to get as few penalty points as possible within the time limit. The leader determines the length of time that the challenge match will be. Two or three minutes may be a good time period. Once the time limit is determined, it should remain the same so that players can measure their improvement.

Modifications

- Ask players to try to use only their nondominant foot (the one not usually used). This modification can be given to the players as a challenge and does not have to be monitored. You might want to reduce the number of balls when you originally add this challenge.
- Place several empty plastic bottles (one-liter soda bottles) in the center to be knocked over or to be avoided.
- Partner Frantic Ball: If you have some "serious" players who want to keep score, have them hit a single tennis ball back and forth. They can record their highest continuous score without a "lift" and then keep trying to better their best score. Keeping score may not be necessary; some players may want to keep score, and others may not.
- Triad or Quad Frantic Ball: Playing in groups of three or four may encourage directional control. Using an occasional "reverse" call may also help players learn directional control.

Comments and Suggestions

- Time is often wasted with lots of questions about how to play. If your instructions have been clear, this is simply a repeat of what you have already said. To allow for more action, you should provide a brief explanation of the game, and then indicate to the group that the first few rounds of the activity will be played "just for fun" so that the players can learn by doing. Encourage players to assist each other in understanding the game.
- Stopped balls can be given a count of 10 by the official before a penalty is given.
- You can add or remove balls if you think it would help the activity. Observation is the best way to determine when balls should be increased or decreased in order to help meet your objectives for the group and individuals. Can you see why *too many* balls can lead to true chaos?
- If your group goes wild in this game, you can reduce the number of balls, or you can consider trying Partner Frantic Ball as described under Modifications.
- You should encourage the use of both the left and right foot.
- Although players start in a circle, as the game progresses, a random formation may evolve. If this works, let it happen. If the situation becomes too chaotic, you might use a circle formation and indicate that any ball going out of the circle is "gone" and should not be recovered. Stop the game when there are not enough balls to be a challenge to the group. Ask the group "Would you like to try again?" If control is still lacking, consider using Partner Frantic Ball. You might also want to use Partner Frantic Ball as a lead-up to Frantic Ball.
- Because you may need a lot of balls, you might want to ask your tennis-playing friends or local tennis club for old tennis balls. If you have a local tennis ball factory or sporting firm, seek a donation of rejects (the balls used in this game do not have to bounce or be "alive").
- By distributing old tennis balls, you may encourage players to do this activity at home.
- Here are eight reasons for using tennis balls rather than soccer balls:
    1. Availability (cheaper, easier to come by)
    2. Encourages better focus (players are more apt to concentrate on a smaller ball)
    3. Leads to a better contact point
    4. Provides less deflection of balls with each other
    5. Encourages the use of the inside (larger part) of the foot and leads to a good soccer technique
    6. Reduces injury potential as compared to a soccer ball
    7. May encourage players to do the activity at home
    8. Encourages flattening of the swinging arc

All these factors support increasing skill development.

- By stressing keeping the ball on the ground, this game encourages a good follow-through, which supports flattening of the swinging arc of the foot. This increases the chances of the ball being contacted behind rather than underneath. Encourage the players to "push" rather than "hit" the ball.
- It might be helpful to ask players what they believe causes the ball to lift off the floor. The answer is hitting underneath the ball, which is probably the result of not flattening the swinging arc.

- Regular soccer balls or playground balls can hurt since they are more apt to be lifted from the floor which can lead to injuries.
- As a warm-up or a challenge, you might ask each player to take a tennis ball and see how many times she can hit the ball against the wall without lifting it. If you do not have sufficient wall space or tennis balls for this, you could have the players choose partners and hit back and forth (Partner Frantic Ball).
- Because some of the growth available in this game can be met only through experience (effective individual planning, group strategy, and so on), you may want to include this activity in your program several times over an extended period of time.
- Loss of equipment is a frequent problem in all play, but having sufficient balls for Frantic Ball is vital. At the end of the game, ask participants to help you find all the balls. Then have each participant drop her ball into the bag. It seems too simple, but it really works.
- This activity enables players to practice all six visual evaluation skills covered in chapter 9. These skills help people read their environment in preparation to solve a problem.

Freeze

Body control is key to every sport and can be improved through play. The game Freeze will help players develop body control.

Equipment Needed

None

Mechanical Principles Being Practiced

- Balance control when making quick stops and maintaining a challenging position
- Absorbing or controlling force
- Kinesthetic awareness of a loss of balance (this awareness can occur in time to avoid a fall by readjusting the center of gravity over the base of support)

Instructions

The players begin to move, slowly at first. Each player challenges himself with different movements. As the game progresses, the players are encouraged to increase activity and the type of movements they are using. At the signal "freeze," all players try to become immediately still and hold the frozen position until "melt" is called, at which time they begin to move again.

Modifications

- Have players do this activity in slow motion. This makes balance more difficult.
- Have players do the activity with partners; one moves in slow motion, while the other calls the signals of "freeze" and "melt."
- Play music. When the music stops, the players freeze. "Melt" occurs when the music begins again.
- Call out the following challenge for any player who wishes to accept it: "Can you hold your position with one or both eyes closed?" This increases the balance challenge and kinesthetic awareness.

Comments and Suggestions

- Body control and balance are important skills in play, sports, and life.
- If the playing area is small, Freeze can be played without players moving from their spot or personal space. Encourage players to take difficult balance positions, including balancing on one foot. They should be sure not to favor one foot.

Golf Tee Challenge

The Golf Tee Challenge is great for improving drives in golf, but it can also improve a player's skills in hockey, soccer, batting, tennis, or any other sport that involves hitting or kicking an object. Appendix A (Putting It All Together: Some Suggestions on Striking) contains useful information that is relevant to this game.

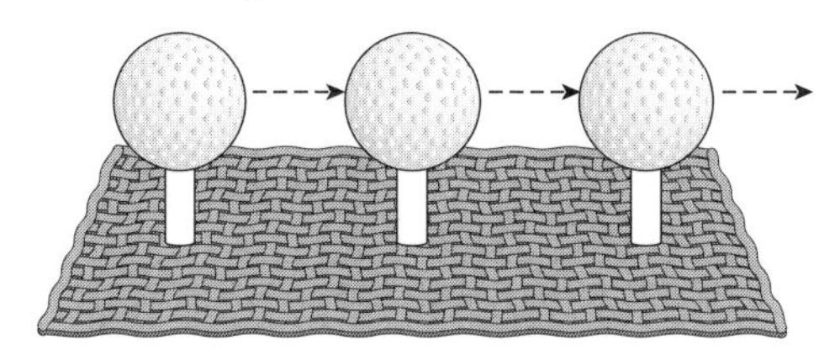

Golf Tee Challenge practice will help a golfer to flatten the swinging arc, improving both distance and accuracy.

Equipment Needed

- Three rubber practice tees
- A doormat that tees can be pushed through
- Lots of plastic practice balls
- A sheet to be hung and hit into (optional)

Mechanical Principles Being Practiced

- Increasing sensitivity to the effects of contact or impact points
- Increasing sensitivity to the effects of the angles of an impact surface
- Flattening of the swinging arc by bending the forward knee, transferring weight, sequentially timing the use of each involved body segment, and using a good follow-through to hit the last ball. This supports the development of increased force and accuracy.
- Follow-through (especially as the number of tees increases from two to three)
- Balance

Instructions

Challenge the golfer to see if she can hit two balls off two tees with one swing. Next, have the golfer try three tees. Initially, the tees can be lined up close to each other. As the golfer becomes more skillful, the tees can be moved farther apart. If practicing at home, a golfer can hit into a sheet or blanket hung over a rope line in the backyard, basement, or garage.

This challenge encourages players to keep their eyes on the ball. It also encourages them to use a sequential swing involving all appropriate body parts, a bent forward knee on the forward swing, a good weight transfer, and a good follow-through. These actions assist the hitters in flattening the swinging arc, which helps them develop increased force and accuracy.

Comments and Suggestions

- Flattening of the swinging arc can lead to greater force and accuracy in many sport skills. Participants can transfer the components involved in the flattening of the arc to tennis, batting, bowling, soccer, hockey, and so on.
- By practicing with multiple golf tees, the golfer is habituating the body to automatically perform the mechanical principles that are common to all sport skills that involve flattening of the swinging arc.
- Having more than one teed ball to hit also encourages focus and concentration.

Living Basketball

Living Basketball is a game that enables players to continue developing their awareness of the parabolic path allowing them to increase their ability to send an object where they want it to go and know where an object will come down (see pages 65-66). This skill plays an important role in all ball sports, from basketball to soccer to golf.

Equipment Needed

- Several large trash bags (depending on the size of the group)
- Several softball-size *soft* objects such as paper-packed balls made by loosely wadding newspaper into a ball and securing it with rubber bands or tape. Because of its force-absorbing quality, this type of ball is safer for the bag holders. Players can make their own equipment.

Mechanical Principles Being Practiced

- Reading the parabolic path (for both thrower and catcher)
- Following through
- Absorbing force (for the bag holders)

Instructions

In this collaborative activity, everyone wins. Teams can have three or four members. The small team size increases the number of turns available to each player. One or two members of the team hold open a large trash bag. The game challenge is to put as many balls into the team bag as possible. Bag holders (who can move to try to catch a ball in the bag) and shooters can both participate in making a goal as they use their awareness of the parabolic path to anticipate and predict the path of the object. The balls can be removed from the bag at any time to allow for more turns, or the bottom can be cut out so that the balls continuously roll out of the bottom of the bag. Rotate the shooters and bag holders.

Comments and Suggestions

- If young players are not yet able to hold the bag (basket) for others, you may consider using a large box. This can also be used when there is only one player.
- This is a good game for players to play at home.
- Encouraging players to make their own equipment at home may increase the amount of practice and may thus improve the skills and mechanical intuition being developed.
- This activity enables players to practice all six visual evaluation skills covered in chapter 9. These skills help people read their environment in preparation to solve a problem.

Partner Scoop Play

In Partner Scoop Play, tossing and catching an object using a gallon jug scoop practices the mechanical principles found in flattening the swinging arc, following through, reading the parabolic path, and absorbing force.

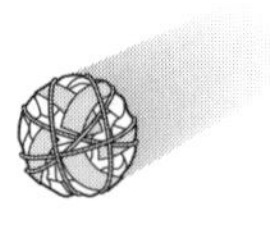

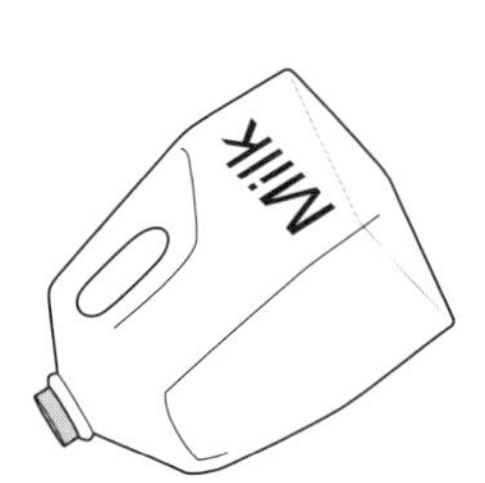

Equipment Needed

- Clean plastic gallon milk, juice, or water jugs
- A sharp knife or scissors to cut out the bottom of the jugs
- Tape to cover the cut edges of the scoop
- A small ball such as a newspaper-packed ball for each set of partners
- Tape and rubber bands to secure the newspaper balls

Mechanical Principles Being Practiced

- Flattening of the swinging arc in finding how to toss the ball successfully out of the scoop
- Follow-through
- Reading the parabolic path
- Force control and absorbing force when catching

Instructions

Players can start by making their own equipment. To do this, they cut out the bottom of the gallon milk jug and tape the cut edges for additional safety. They now have a scoop catcher. To create the ball, players firmly wad newspaper into a ball and secure it with rubber bands and tape.

Partners play catch using the scoop to both throw and catch the ball.

The scoop must be manipulated in a certain way to get the ball out of the scoop for a successful toss and to successfully catch the ball without it bouncing out of the scoop. Because of this, the game is helpful in increasing the players' awareness of the mechanical principles involved in flattening the swinging arc, following through, and absorbing force. This activity also practices reading the parabolic path.

Comments and Suggestions

- When collecting milk or juice jugs, you should wash the jugs thoroughly immediately after emptying them and then cut out the bottom. This reduces smells and stickiness, and it makes stacking possible, reducing the storage space needed.

- The position of the opening of the scoop tends to encourage flattening of the swinging arc and a good follow-through. This allows the ball to be released in a forward rather than upward direction. Players usually discover this for themselves.
- Players learn through experience that drawing the scoop backward when receiving the thrown object tends to absorb force and keeps the object in the scoop. This is helpful when they later receive heavier or harder or faster thrown objects in sports.
- This activity enables players to practice all six visual evaluation skills covered in chapter 9. These skills help people read their environment in preparation to solve a problem.
- Encouraging players to make scoops and newspaper-packed balls at home (for themselves and others) may increase the number of throws and catches being practiced and the mechanical intuition and skills being developed.
- This is a good activity for the players to do at home.

Pass-a-Puck

Because this activity encourages a flattened flinging motion to propel the puck, the activity can support future success in many sports, such as hockey, soccer, golf, and bowling.

Equipment Needed

- Hockey sticks (see appendix B for information on how to make your own, page 131)
- Sturdy canvas-covered, button-filled beanbags (buttons make a good filling; they are much less apt to deteriorate). See appendix B for information on how to make your own, page 131. If the playing space is carpeted, you may also need Ziploc bags (putting the button bags inside of these will help them slide).
- Something to indicate a center dividing line
- If using the modification with skates, you will need pieces of carpet, or rags (approximately 6 by 11 inches [15 by 28 cm]) to act as skates. As an alternative, large socks could be pulled over players' shoes. See appendix B for more information on skates. If the space is carpeted, cardboard, shoe boxes, shoe box lids, or paper plates can be used as skates.

Mechanical Principles Being Practiced

- Balance (abdominal and leg strength and endurance, plus kinesthetic awareness, especially when the game is played on skates)
- Flattening of the swinging arc in flinging the button-filled beanbag puck
- Follow-through

Instructions

Initially, Pass-a-Puck can be played without skates under players' feet. Divide the group into partners. If you have an extra player, have one group of three.

Players practice passing the puck back and forth over a short distance using a flinging motion. Instruct them to put their stick on the button bag (puck) and, without lifting it from the bag, fling the bag to one another. This leads to a flattening of the swinging arc and a good follow-through. The flinging action also helps to prevent potential injuries to other players that could be caused by dangerous backswings.

Encourage the players to move farther apart as they find ways to increase the distance they can fling the button bag. When players have had time to gain some skill in this flinging technique, divide the group in half. If they have been working in partners, simply have each partner go to one side of the center line. Give plenty of pucks to each side (team). On a signal, all the players attempt to fling all the pucks on their side onto the other team's side; this includes any button bags sent onto their side by the other team. The action continues until the signal is given to stop. This is so much fun that scoring is usually unnecessary. If players insist on keeping score, the button bags on each team's side at the end of the round can be counted. The team with the fewest button bags on their side wins that round. We have found that keeping score tends to result in several button bags being flung to the opponent's side after the signal to stop has been called. If some players want to know who won, you can do as a colleague does. He asks the players who had fun or who got better at flinging the puck. He waits until the players raise their hands, and then he says, "You did [won]."

Ask players if they would like to try again. Later consider putting the players on skates (see the upcoming modification).

Modification

- Have the players use "skates" (see the activity of Skates on pages 117-118).

Comments and Suggestions

- Button bag pucks compress on impact to absorb the force, thus reducing potential injuries.
- Heavy duty covers on button bags are important for durability. Buttons with smooth edges are especially good to use as filling. Beans can deteriorate over time and with use. A friend uses an old glove filled with buttons as a puck.
- Button bag pucks tend to encourage putting the stick on top and using a flinging motion. This leads to a more skillful pass that includes a flattening of the swinging arc and an extended follow-through, which is a good lead-up technique for both ice and field hockey. It also reduces the tendency to backswing wildly.
- For real durability in your homemade hockey sticks, put 3/4-inch nonflexible PVC (plumbing) tubing inside the rolled newspaper covering. Sticks wrapped in several layers of newspaper absorb force on impact with other players. This can reduce injuries that could result from the use of wood or plastic sticks.
- Limiting the length of the stick to the length or width of a full sheet of newspaper causes players to bend their knees to fling the puck. This strengthens the bent-knee position, which is important in maintaining one's balance and the ready position in sports. Bending the knees also contributes to the flattening of the swinging arc and the follow-through, and it is a valuable factor in many games and sports, even in bowling and the underhand pitch in softball.
- Covering the ends of the sticks with a sturdy tape will help them last longer and will keep the hard core within the stick.
- Encouraging players to make their own equipment (for themselves and others) at home may increase the amount of practice and may thus improve the mechanical intuition and skills being developed.
- This activity is a good lead-up to Rag or Rug Hockey. In addition, the equipment made for Pass-a-Puck can be used in Rag or Rug Hockey.
- This activity enables players to practice all six visual evaluation skills covered in chapter 9. These skills help people read their environment in preparation to solve a problem.

Racket and Balloon

Hitting—whether it's with a foot, hand, racket, stick, bat, or club—always involves the same basic principles. Learning this when young can prepare a player for success in many sports. Activities such as Racket and Balloon support the development of visual evaluation skills. These skills include focus and concentration, tracking, reading and predicting the movement of an object, figure and ground discrimination, and judging distance and speed. (See pages 93-95 for more information on these visual evaluation skills.)

Equipment Needed

- Strong (sturdy) wire coat hanger
- One leg of ladies stretch hose or some thin spandex type cloth
- Masking tape
- Scissors
- A pair of pliers (may be needed to shape the handle of the racket)
- A balloon or two for each individual or set of partners

See appendix B (page 132) for instructions on how to make a racket for this activity. See also Torbert's article "Growing as One Plays With a Balloon" in the January/February 1990 issue of the journal *Teaching Elementary Physical Education.*

Mechanical Principles Being Practiced

- Awareness of the effect of contact or impact points
- Flattening of the swinging arc
- Angles of the striking surface
- Follow-through

Instructions

Partners can hit the balloon back and forth as in badminton, or a single player can volley the balloon to herself or against a wall.

Comments and Suggestions

- It is always good to have an extra balloon or two.
- Broken balloons can be a hazard.
- Balloons that are blown up only to two-thirds or three-fourths their full capacity are much less apt to break. Consider putting a balloon inside a balloon and blowing up only the inside balloon (double balloon).
- Encouraging players to make their own equipment (for themselves and others) and play at home may increase the amount of practice and may thus improve the mechanical intuition and skills being developed.
- Homemade equipment can be made and given as gifts by the players.
- Racket and Balloon is a good activity for players to play at home.
- This is a wonderful intergenerational activity.

Rag or Rug Hockey

Games such as Rag or Rug Hockey can help players improve body control, balance, coordination, and the lower body endurance needed in many sports. Players will also learn teamwork where every player is important. Pass-a-Puck (pages 111-112) is an excellent lead-up to this game. Plus, you can use a lot of the same equipment.

Equipment Needed

- Pieces of carpet, or rags (approximately 6 by 11 inches [15 by 28 cm]). As an alternative, sock skates (large socks pulled over shoes) could be used. If the space is carpeted, use cardboard, shoe boxes, shoe box lids, or paper plates to act as skates.
- Some larger pieces of carpet or rags to act as "twisters" (optional; see the description in the Instructions section)
- Newspaper-wrapped (to absorb force) sticks with a PVC or other sturdy core to retain rigidity
- Sturdy canvas-covered, button-filled beanbags (buttons make a good filling; they are much less apt to deteriorate). See appendix B (page 131) for information on how to make your own. If the playing space is carpeted, consider putting the button bag in a plastic Ziploc bag. This will allow the puck to slide better.
- Something to indicate a center line
- Cones or plastic bottles with some ballast in them for goal markers
- Strips of cloth (approximately 1 by 24 inches [2.5 by 61 cm]) to be tied around players' legs in order to identify one team—torn-up bed sheet is a possibility.

Mechanical Principles Being Practiced

- Balance (leg and abdominal strength and endurance; kinesthetic awareness)
- Flattening of the swinging arc in flinging the button bag
- Follow-through

Instructions

Rag or Rug Hockey is played like ice or floor hockey, but the players use special "skates" (e.g., a piece of carpet, rag, or cardboard). The game is played with as few rules as possible; however, no body checking (intentionally knocking another player out of the way) is allowed. The players can be encouraged to make up rules as they see a need for them.

Each participant stands with each foot on a skate. Test out various sizes and evaluate which works best for the individuals in your group. To move, a player must use a cross-country skiing or skating motion. Players must keep a skate under each foot at all times. If a player loses a skate, she cannot participate in the action until the skate is back under her foot.

Offer players the choice of who will be the goalies and for how long. Goalies cannot use their hands. The playing area can be determined by the space available, and how far apart the goal markers should be placed can be a decision of the two teams.

When there are two teams of unequal ability, a team that is two points behind designates one member of the *opposing* team as a "twister." This player must put both feet on only one larger rag or rug and must twist to move. This has proven to be very complimentary to the one chosen by the opposing team and is called "complimentary handicapping." This twister player may now begin to use abilities that she has not had to use previously (strategies, teamwork, and so on). If a team gets four points behind, a second twister can be designated. If the losing team begins to catch up, every time this team gains two points on the opponent, one twister returns to her original skates. For three valuable add-on rules that have been created by players, check the first three items under Modifications.

If your players have been using socks, rags, carpets, shoe boxes, or cardboard as skates (see the Skates activity on pages 117-118), they have probably already developed the skills required to keep the skates under their feet. As a result, they will be ready to do this in Rag or Rug Hockey with less frustration.

Modifications

- An 11-year-old suggested giving the selected player the choice of becoming a "twister" or playing for the other team until the two-point deficit is reduced.
- A 7-year-old recommended a "no stealing the puck" rule. When a player has the puck under his stick and is not moving it, all other players must stay at least 4 feet (122 cm) back from this player. This seems to work well. We have found that this also reduces the bunching that can occur in team sports.
- An adult player suggested this rule: When the goalie has the puck, everyone must stay back 6 feet (183 cm) to allow the goalie to fling the button bag puck away from the goal.
- Participants in wheelchairs or leg braces can play without rugs or rags under their feet.
- With some groups, we introduce a "no dribbling" rule (passing only) to encourage more team play.
- One-on-One Rag or Rug Hockey: This modification can be used to provide opportunities for one-on-one play. The experience could also plant the seed for play at home. The use of a cardboard box as a goal enables players to design their own versions of the game from homemade materials.
- Small groups can play with only one goal.

Comments and Suggestions

- The use of "skates" is especially valuable when dealing with large groups or limited space. Moving on these skates produces a motion that is vigorous for the participants while being much slower than the traditional running in floor hockey. The slower, but vigorous, movement results in decreased injuries because the force of any player impact is reduced. This might also be a good activity for an ice hockey team when ice is not available or when practice time is too expensive; a field hockey team could use this activity inside when it is raining.
- Heavy duty covers on button bags are important for durability. Buttons with smooth edges are especially good to use as filling. Beans can deteriorate over time and with use. A friend uses a glove filled with buttons as a puck.
- Button bag pucks compress on impact to absorb the force, thus reducing potential injuries.
- Button bag pucks tend to encourage putting the stick on top and using a flinging motion. This leads to a more skillful pass that includes a flattening of the swinging arc and an extended follow-through, which is a good lead-up technique for both ice and field hockey. It also reduces the tendency to backswing wildly (which can be dangerous).
- The newspaper-wrapped stick gives to absorb force. This can reduce injuries that could result from wood or plastic sticks.

- Limiting the length of the stick to the length or width of a full sheet of newspaper causes the players to bend their knees to fling the puck. This strengthens the muscles involved in the bent-knee position. This is important in maintaining one's balance, flattening the swinging arc, and performing a proper follow-through, which are valuable skills in many games and sports.
- Covering the ends of the sticks with a sturdy tape will help them last longer and will keep the hard core within the stick.
- Encouraging players to make their own equipment may lead to additional play in other places and may thus increase the development of mechanical intuition and related skills.
- Players usually need some practice using rag or rug skates before they are ready to play Rag or Rug Hockey. Let the participants skate around the playing area for a while, or consider playing Skates (pages 117-118) and Pass-a-Puck (pages 111-112). Pass-a-Puck can provide practice in both skating and flinging of the puck.
- Tying a strip of material (perhaps from a torn-up bed sheet) around the leg (just below the knee) of each member of one team allows the teams to be differentiated. This strip should not be knotted or tied too tight or too loose.
- This activity enables players to practice all six visual evaluation skills covered in chapter 9. These skills help people read their environment in preparation to solve a problem.

Rebound Board

The rebound effect is determined by the same basic mechanical principles whether it occurs in basketball, soccer, golf, or tennis. The player's objective is to send an object in a specific path.

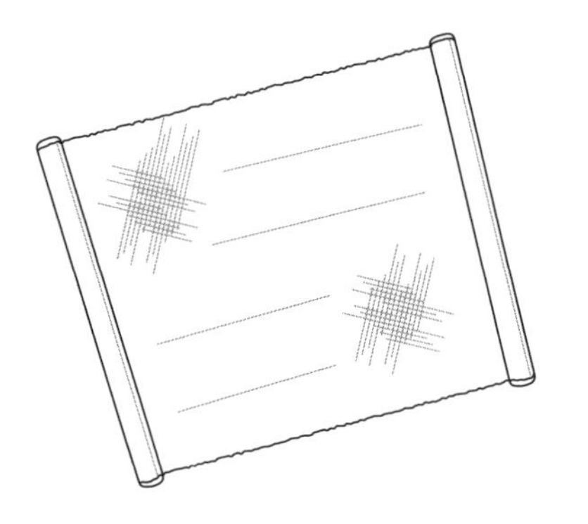

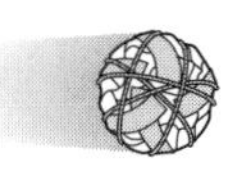

Equipment Needed

- Stretchy material (a small T-shirt may be used). Refer to appendix B (page 133) for information on how to make a rebound board from stretch material.
- Two 10-inch (25 cm) sticks for each rebound board. These can be made of firmly wrapped newspaper.
- A firm surface such as a three-ring notebook (optional; this can be used instead of a rebound board made of stretch material)
- Small paper-packed balls made by wadding newspaper into a tight ball and securing it with rubber bands or tape

Mechanical Principles Being Practiced

- Flattening the movement at contact (a pushing motion)
- Awareness of the effect of contact or impact points and angles of the striking surface
- Follow-through
- Reading the parabolic path

Instructions

Have one player toss a small paper-packed ball toward the rebound board. The player holding the rebound board attempts to direct the object back to the thrower. Because a paper-packed ball will absorb some of the force, it will travel slower off the rebound board surface, allowing the players to have a little more time to read the effects of the mechanical principles.

Comments and Suggestions

- This is a good at-home activity for players. It can be a fairly quiet activity and may be less apt to bother people in the vicinity.
- Encouraging players to make their own equipment (for themselves and others) at home may increase the amount of practice and may thus improve the mechanical intuition and visual evaluation skills being developed.
- This activity enables players to practice all six visual evaluation skills covered in chapter 9. These skills help people read their environment in preparation to solve a problem.

Skates

This activity is good for balance and abdominal development, which help players play better in all sports.

Equipment Needed

- Pieces of carpet, rags (approximately 6 by 11 inches [15 by 28 cm])
- Alternative—socks for pulling over feet or shoes
- If the playing space is carpeted, cardboard, shoe boxes, shoe box lids, or paper plates can be used as skates.

Mechanical Principles Being Practiced

- Balance and total body coordination (kinesthetic awareness; abdominal development and leg strength and endurance for supporting the center of gravity over the base of support)
- Force development (abdominal development for pelvic stabilization)
- Accuracy (abdominal development for pelvic stabilization)

Instructions

Each participant places a skate under each foot. To move, a participant uses a cross-country skiing or skating motion. The players can now play all kinds of games or challenges on these skates.

Modifications

- Sock Skates: If the use of these skates under the feet is too difficult for players, try having the players wear large socks over their shoes. Some floors may be too slippery for this modification.
- Scooter Skating: If players have difficulty dealing with skates under both feet, consider using a skate under just one foot or a sock on only one foot. Have the skaters use a scooterlike movement pattern. Challenge players to use a different foot as the scooter at different times.

Comments and Suggestions

- Putting players on "skates" allows more players to participate in less space, and it also allows for vigorous activity with reduced potential for injuries because of the lower impact force if collisions occur.
- Some participants have used shoe boxes (tops and bottoms) as skates. Look around. What's available?
- Players with a wide range of ability can function effectively together when playing on skates.
- Note that Pass-a-Puck (pages 111-112) and Rag or Rug Hockey (pages 114-116) are both played on skates. Freeze (pages 107-108) can also be played on skates.
- All activities played on skates develop the abdominal muscles that control the pelvic region, which is used to adjust the center of gravity over the base of support to maintain balance and stabilize the hips. This provides greater force production in sports such as golf, soccer, tennis, and gymnastics. Playing on skates also helps develop the proprioceptive or kinesthetic system to alert a player if she becomes off balance. This is important for a rapid adjustment.
- The activity Skates might be considered when ice or field time is not available for hockey.
- Encourage players to make their own skates for home use. This is a good at-home activity because these types of skates are quiet and can be used in less space.
- Refer back to chapter 2, pages 20-21, for how Skates helped a young boy improve his balance.

Note: See also Torbert and Stork's article "'Skating' and 'Skiing' on Special 'Skates' and 'Skis'" in the March 2006 issue of *Teaching Elementary Physical Education.*

Super Sox

The force-receiving position needed in so many sports is a natural outcome of Super Sox.

Equipment Needed

- Two large tube or athletic socks for every four players (socks must fit easily over the largest arms in the group)

Mechanical Principles Being Practiced

- Balance and total body coordination (kinesthetic awareness; abdominal development and leg strength and endurance for supporting the center of gravity over the base of support)
- Taking a natural force-absorbing stance (forward and backward stride) and bending the knees to absorb force as the socks are pushed onto the arms. (This is the same stance that a person would take to effectively receive and absorb any oncoming force. By becoming habituated to these force-absorbing techniques, participants can apply them in other situations.)

Instructions

Create teams of four. If your group doesn't divide evenly by four, the extra players can coach their team and rotate in after each round or practice trial.

One player puts the tube socks on over his hands and up his arms. Another teammate stands in front of this player—these two players have their hands positioned fingertips to fingertips with thumbs tucked in so the socks won't get hung up on them. The other two players each take hold of one sock and, turning it inside out, pull it off the first player and onto the arms of the second player. The second player then turns to the third player. Players one and four pull the socks onto the third player and so on with the fourth player receiving the socks next. This process continues, and each group counts how many changes of socks they can do in one minute.

Technique and team strategy are very important. Let the teams practice before being timed. Time all groups at once. Each team counts its own changes. Teams repeat this process, trying to increase the number of changes of socks they can do each time.

Modification

Hand Wrestling: See the description in chapter 2 (page 19).

True or False Review Questions

Information about each question is found on the page number provided within the parentheses following the question.

REMEMBER THAT A PARTIALLY FALSE QUESTION IS CONSIDERED FALSE.

Answers to the questions and additional information are found in appendix C.

1. The games in this chapter can be used for some end-of-practice fun or as family activities. Their function is to enable players to practice developing sport skills (pages 97-98).
2. The parabolic path refers to the mirror-image path taken by an object in flight. Various forms of air resistance can change this path. Although the awareness of the parabolic path allows a player to predict the usual path of an object, air resistance may require adjustment (page 65).
3. The games in this chapter are meant to help teachers, coaches, and parents explain the various mechanical principles of movement so that a player can apply these principles in play and sports (pages 97-98).
4. The mechanical principles of balance, force production, motion, and leverage are identical regardless of the activity (page 4).
5. Learning to flatten the swinging arc can be transferred into sports such as tennis, golf, bowling, soccer, hockey, and diving (pages 99, 109).
6. Blanketball involves reading the effects of the angle of the force-producing surface, reading the parabolic path, and absorbing force by giving with the force (page 102).
7. In Frantic Ball, starting with just one ball allows the group to get the idea of what is to be done and helps prevent chaos from developing (page 105).
8. In Frantic Ball, the emphasis on keeping the balls on the ground supports flattening of the swinging arc, and flattening the swinging arc will help keep the balls on the ground (pages 105, 106).
9. Kinesthetic awareness of a loss of balance can occur in time to avoid a fall by readjusting the center of gravity over the base of support (page 107).
10. In Living Basketball, both the shooters and the bag holders are learning to read the parabolic path to make predictions, and the bag holders are also using the absorption of force (page 109).
11. In Partner Scoop Play, tossing an object using a gallon jug scoop may be more helpful than playing regular toss and catch when it comes to increasing the use of the mechanical principles involved in flattening the swinging arc, following through, reading the parabolic path, and absorbing force (pages 110-111).
12. Rag or Rug Hockey is a good lead-up to Pass-a-Puck (page 112).
13. Although Racket and Balloon play involves four important mechanical principles of movement (contact points, flattening the swinging arc, angle of the striking surface, and follow-through), only three of these mechanical principles (contact points, angle of the striking surface, and follow-through) will transfer to tennis (page 113).
14. When two teams are of unequal ability in Rag or Rug Hockey, a team that is two points ahead must choose one of their team members to play as a "twister" (page 114).

15. Putting players on "skates" in Rag or Rug Hockey will not reduce the impact force if collisions occur, but it can allow you to have more players participating vigorously at any one time and may allow more activity in less space (page 115).

16. In Rebound Board, players are practicing only the important mechanical principles related to contact points and angles of the striking surface. The rebound effect in relation to these is the same as found in golf and tennis (page 117).

17. If the area is carpeted, participants can use cardboard, shoe boxes, shoe box lids, or paper plates as skates (page 117).

18. The activity Skates is good for abdominal development, which supports pelvic stabilization for force development and accuracy and supports balance by maintaining and adjusting the center of gravity over the base of support. Both are important in most sports (pages 117, 118).

19. Playing on "skates" develops the abdominal muscles, which help adjust the center of gravity over the base of support to maintain balance, but the abdominal muscles are not involved in the force production needed in sports such as golf (pages 31, 118).

20. Super Sox supports balance, absorption of force, and follow-through (page 119).

21. In Super Sox, the bent-knee stride position normally taken by the person receiving the socks is good in relation to absorbing force and maintaining balance. This is a common position taken when receiving a strong pass in basketball (page 119).

PUTTING IT ALL TOGETHER
Some Suggestions on Striking

This appendix demonstrates how you might use the materials in this book to reach your specific goals. Because striking is an important skill in many sports, it will be used as an example.

As you continue to read *Secrets to Success in Sport & Play* and use an analytical approach, you will find yourself more and more able to make a list of general and specific needs for a particular skill. Much like making a grocery list, this will help you organize. Each of you will develop your list in your own personal way.

Doing a needs analysis is an evolving process. Continue to evaluate as you work on any specific skill. Don't forget that there are preparation or awareness, action, and completion or follow-through stages.

The page numbers listed after the various factors indicate where you can find information in this book related to that factor. This is helpful if you want to review related material.

General Needs Related to Striking

- Relax so you can get a full swing that is free and easy. This will not only add force but will make the path of your swing more consistent (page 34).
- Keep your eye on the object to be struck (page 93).
- Predict the path of the object being sent and received. (Know the effects of gravity [pages 76-77], the parabolic path [pages 65-66], air resistance [page 77], spins [pages 58-61 and 66-67], and rebounds [pages 57-59].) Don't wait for the ball to get to you before beginning to determine its path.
- Swing level or in line with the path of the object to be struck so you can meet the ball at any of several points along its path (figure 1.2 on page 7).
- Flatten the swinging arc for both force and accuracy (pages 31-32, 77-80).
- Be sure you have sufficient strength in all involved muscles in order to stabilize all contributing body parts for both accuracy and force (pages 29-31).
- Will you be ready for the next move?

Specific Needs Related to Striking

If you need force, you should do the following:

1. Use a long lever—one that you can still control (page 40).
2. Study factors that help you overcome inertia (pages 25-26, 27).
3. Take a complete backswing (pages 25, 29, 32 [full stretch] and page 33 [range of motion]).
4. Involve the stretch reflex (pages 25, 31).
5. Be sure all body parts that could contribute are involved (page 29) and that each has appropriate strength (pages 29, 30) and range of motion (page 33). Include trunk rotation (pages 29, 32) and opposition (page 39).
6. Add body parts sequentially from the center of gravity out to the end of the involved levers (page 30 and figure 4.2 on page 30).
7. Be sure you are stabilizing previously involved parts of the body (pages 29, 30, 31). Be careful that you do not "give" on impact.
8. Transfer weight in the desired direction (pages 29, 31, 32) in the preparatory, action, and follow-through phases.
9. Follow through (pages 29, 33, 37, 39).
10. Increase the time and distance over which your force is developed (page 32; you might also want to note pages 31-32 [flattening the swinging arc], page 33 [increase the range of motion], page 39 [use opposition], page 32 [trunk rotation], page 32 [lengthening levers], and page 32 [weight transfer]).

If you need accuracy, you should do the following:

1. Shorten the lever for greater control (page 40).
2. Flatten the swinging arc so you can hit at several points in the path of the swing (pages 31-32, 78; also see figures 8.9 and 8.10).
3. Be sure your swing and follow-through are in the desired direction (pages 36, 79, 80).
4. Understand the effects of the direction of the applied impact force (pages 35, 80), the contact point (pages 80-81), the angle of the impact surface (page 81), spins (pages 58-61 and pages 66-68), roll and rebound patterns (pages 57-61), the parabolic path of a projectile (page 65), and follow-through (page 83).
5. Remember that balance (pages 13, 80, 82) and stabilization (page 82) are also important in accuracy.

If you need speed of rotation of the striking implement in order to get quickly to the object to be struck, you should do the following:

1. Shorten the striking lever (pages 30, 41).

2. Shorten the preliminary actions—reduce trunk rotation or eliminate the backswing (page 30).
3. Remember that opposition may become irrelevant (page 30).
4. Understand that each body part can make less or no contribution depending on the involved needs (page 30).

APPENDIX B

HOW TO MAKE EQUIPMENT

Why use homemade equipment? When I was growing up, my family was poor, and we often made our own equipment. We found that making homemade "stuff" was fun, plus it allowed us to play just like all the other kids. In addition, while those other kids were losing turns when their purchased equipment sprung a leak or went down the gutter into the sewer, we were still playing and improving our skills.

We were able to make things for our friends or show them how to make things. Making our own things seemed to allow for a certain creative magic. When we didn't have something, we figured out how it could be made with what we did have. We also created challenges such as the Batting (or Golf) Tee Challenge described in chapter 10 (pages 101 and 108). We could play games with newspaper-packed balls and not have to worry about younger siblings getting hurt. And for the kids who weren't so good yet, a paper-packed ball was easier for them to catch.

This appendix describes how to make various types of equipment, including newspaper-packed balls, batting tees, scoops, rackets (for Racket and Balloon), skates, button-filled beanbag pucks and paper-wrapped sticks (for Pass-a-Puck and Rag or Rug Hockey), and rebound boards. When players make their own equipment, this may actually increase the places they can play and the amount of practice that will occur. Soft homemade equipment such as that mentioned is not only safer, but is also quieter because it will absorb force on impact. Making equipment for themselves or to share with others can also become an at-home activity for players.

Have players do as much of the work as possible in making their own equipment. Others can help, but only when help is really needed. Homemade equipment can be made and given as gifts by the players.

Batting Tees

Before beginning, note alternatives 1 and 2 in this section. You should choose to make one or the other depending on the availability of materials and tools. For further information on the Batting Tee Challenge, go to chapter 10, pages 101-102.

Equipment Needed for Both Alternatives

- A fine blade saw for cutting three-foot lengths of PVC tubing (or you could have them cut at the place of purchase)
- Several softball-size firmly packed newspaper balls made by tightly wadding newspaper into a ball and covering it with tape. Because of its force-absorbing quality, this type of ball is safer, less damaging, and less noisy.
- Bats can be made to the preferred size by using tightly rolled newspaper. Some people like to tape the bat to make it last longer and to prevent newsprint from coming off on the hands. Paper bats compress more, so they absorb the force and thus hurt less if another player is accidentally contacted. The newspaper bat seems to encourage players to focus on the challenge of hitting all the balls rather than how hard the balls can be hit.
- If a batter is playing alone, a sheet or blanket can be suspended from a rope line to hit into. Hits that go only a short distance allow for more practice.

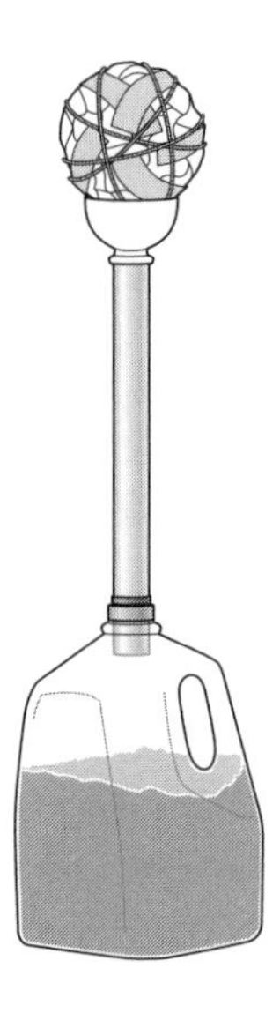

Equipment Needed for Alternative 1

- A one-gallon plastic milk, juice, or water jug
- Two screw-on jug caps
- 3 feet (91 cm) of 1-inch PVC plumbing tubing (1 1/4-inch outside diameter—take an empty jug to the hardware store to test for the best fit)
- The top 2 1/4 inches of a plastic liter soda bottle
- A knife to cut the soda bottle and hole in cap (or use a drill for cap if available)
- A file or rasp (or you can use a rough sidewalk) to scrape the neck of the soda bottle to fit into the upper end of the PVC tubing

- Dirt, rocks, or pebbles to fill at least three-fourths of the jug in order to act as ballast
- A thick rubber band or two

Instructions for Alternative 1

Cut or drill a hole almost 1 1/4 inches in diameter out of the center of one cap. Screw this cap onto the gallon jug to act as a reinforcing ring. It is good to have a second jug cap to use in transporting the disassembled tee.

Put dirt, water, pebbles, or other ballast into the jug to prevent it from tipping over. Put one end of the PVC tubing into the jug. Place a rubber band around the PVC tubing. Moving this rubber band up and down allows for adjustment to different batting heights. Scrape the neck of the cutoff soda bottle top to reduce its size so that it fits snugly into the end of the PVC tubing opposite the jug end. This will act as a cup to hold the ball.

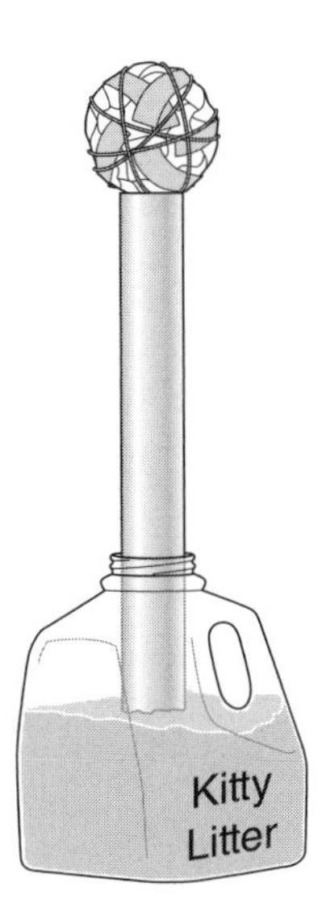

Equipment Needed for Alternative 2

- Large plastic kitty litter container with contents (approximately 15 pounds) and screw-on cap
- PVC pipe of the appropriate diameter to fit into the container opening—take your kitty litter container to the hardware store to test for the best fit (1 1/2-inch pipe has been our choice, but it must fit the opening for your kitty litter container)

Instructions for Alternative 2

Judy Purdy of Eagan, Minnesota, has made the entire process of making a homemade batting tee a great deal easier by using a large plastic kitty litter container. She leaves the kitty litter in the container as ballast. Using the largest diameter PVC tubing that fits into the large opening on the kitty litter container, she simply has the hardware store cut the PVC tubing to the length she wants. The tee can also be made to different heights by simply pushing the PVC tubing to different heights in the kitty litter. Because of the large diameter of the PVC tubing, it will hold the ball without the need for a cuplike addition.

Putting the kitty litter cap back on the container makes storage and transporting relatively easy.

Golf Tees for Golf Tee Challenge

For further information about the Golf Tee Challenge, go to chapter 10, pages 108-109.

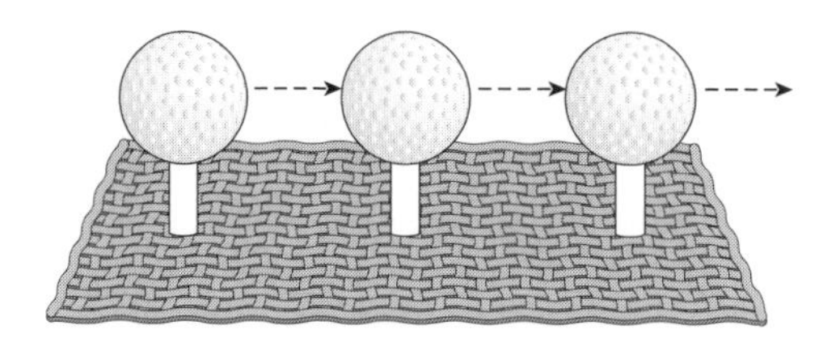

Equipment Needed

- Three rubber practice tees
- A door or golf mat that tees can be pushed through
- Lots of plastic practice balls
- A sheet to be hung and hit into (optional)

Instructions

Push two rubber practice tees up through the mat. When the golfer feels ready, add the third rubber golf tee.

Newspaper Balls

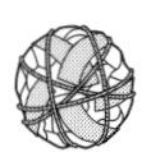

Equipment Needed

- Newspaper
- Rubber bands
- Masking tape
- Plastic bag (optional)

Instructions

A newspaper ball is simply a soft, loosely packed wad of paper about the size of a head of cabbage and held together by one or two rubber bands or masking tape. For longer wear and cleaner hands, you may want to put each newspaper wad in a plastic bag before securing it with tape or rubber bands. The way a ball is made (size and softness or hardness) depends on the particular activity. Newspaper balls are used in Batting Tee Challenge, Blanketball, Chaotic Team Juggle, Living Basketball, Partner Scoop Play, and Rebound Board. Refer to the information for each of these specific games in chapter 10.

Pass-a-Puck and Rag or Rug Hockey Pucks

Equipment Needed

- Buttons
- Sturdy fabric cut into 6-by-12-inch (15 by 30 cm) pieces
- Needle
- Thread

Button-filled beanbags can be used as pucks. Heavy duty covers on the button-filled bags are important for durability. Buttons with smooth edges are especially good to use as filling. Beans can deteriorate over time and with use. Cut the sturdy material into pieces that are roughly 6 by 12 inches. Fold each of these into a 6-by-6-inch pocket and then sew the sides. Turn each one inside out and fill it half full with buttons. Fold the open edges of the bags in and sew the opening shut, or take it to professional seam sewer. Use a sturdy thread and sew at least twice on this final side.

A colleague uses a glove filled with buttons as a puck. The button bag pucks tend to encourage putting the stick on top and using a flinging motion. This leads to a more skillful pass that includes a flattening of the swinging arc, which is a good lead-up technique for ice, floor, and field hockey. It also reduces the tendency to backswing wildly (which is dangerous).

For instructions on how to play Pass-a-Puck and Rag or Rug Hockey, go to pages 111-112 and 114-115 in chapter 10.

Pass-a-Puck and Rag or Rug Hockey Sticks

Equipment Needed

- PVC pipe or other sturdy core
- Newspaper
- Duct tape, filament tape, or other sturdy tape

Instructions

To create your own sticks for Pass-a-Puck and Rag or Rug Hockey, wrap PVC pipe (or some other sturdy core) with several layers of newspaper and then tape the newspaper securely. Covering the ends of the sticks with sturdy duct or filament tape will keep the hard core from falling out of the newspaper. As a result, the sticks will last longer. The newspaper wrapping absorbs the force and thus reduces injuries when ankles are accidently hit. The PVC tubing or other sturdy core retains the rigidity needed in these two activities.

For further information about these sticks, refer to the Comments and Suggestions section for Rag or Rug Hockey in chapter 10, pages 115-116.

Racket and Balloon

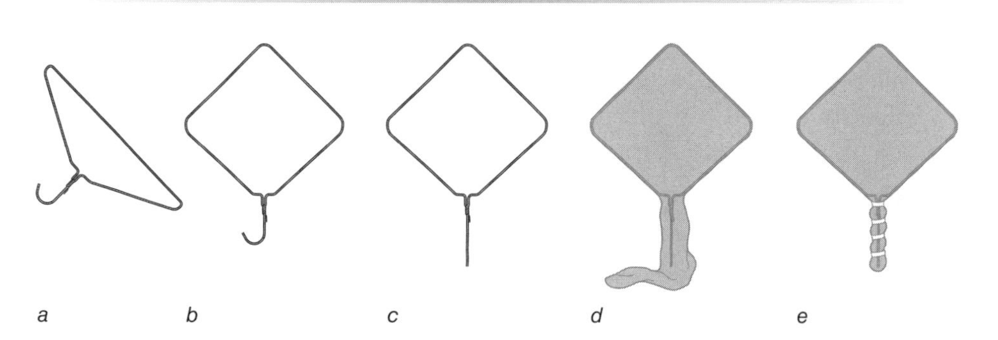

Equipment Needed

- Strong (sturdy) wire coat hanger
- One leg of ladies stretch hose (or some thin spandex type cloth)
- Masking tape
- Scissors
- A pair of pliers (may be needed to shape the handle of the racket)
- A balloon
- A length of ribbon, yarn, or string (optional for a single player)

Instructions

Bend the coat hanger into a diamond or square shape. Bend the hook of the hanger to form a handle. This may require a pair of pliers or some fairly strong hands. Pull the foot of the stocking over the top of the hanger (opposite the handle) and work it down toward the handle. Pull it tight so that a bouncy surface is formed. Wrap the rest of the stocking around the handle to give some padding for the hand, and tape the hose to the handle. Cut off any bulky excess. Using other stretch materials may require some sewing.

Blow a balloon up about two thirds to three quarters of its normal size (reduces breakage). Tie the balloon off securely. Consider putting a balloon inside a balloon and blowing up only the inside balloon (double balloon).

For more information about playing with the racket and balloon, go to page 113 in chapter 10.

Rebound Boards From Stretch Materials

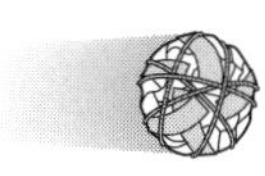

For a stretchy rebound board, place two tightly rolled 10- to 12-inch (25 to 30 cm) paper sticks inside a loop of stretchy material. This could be the seat part of panty hose or a small T-shirt. If you are using material that isn't already formed in a loop like a T-shirt or the seat part of panty hose, cut about 28 inches (71 cm) of material and sew the ends together, creating a loop that will be the size of your rebound board. Newspaper sticks can be made by cutting 20 to 30 strips of newspaper that are 10 to 12 inches tall. Fold the newspaper so the outside (single) edges are together. This creates a fold opposite the single edges. Then roll the single edges toward the fold. Having the folded edge on the outside of the stick tends to prevent shredding. Tape the sticks. Using newspaper rather than wooden sticks helps prevent injury from the stick ends.

Place the sticks inside the loop of material. Holding one stick in each hand, pull the two sticks apart to create the rebound surface.

Rebound Boards From Nonstretch Material

A rebound board with a firm surface can be made from material such as a school notebook, plastic, or lightweight plywood (approximately 11 1/2 by 10 inches [29 by 25 cm]). Sometimes even a heavy piece of cardboard will do.

For more information about playing with rebound boards, go to pages 116-117 in chapter 10.

Scoops

For information on how to make scoops out of plastic gallon milk, water, or juice jugs for Partner Scoop Play, go to pages 110-111 in chapter 10.

Skates

If playing on a smooth floor, each participant should place a piece of carpet or rag (roughly 6 by 11 inches [15 by 28 cm]) under each foot. If playing on carpet, participants should use cardboard, shoe boxes, shoe box lids, or paper plates as skates. To move, a participant uses a cross-country skiing or skating motion. The players can now play all kinds of games or challenges on these "skates." To find out more about using skates, refer to the information about Skates, Rag or Rug Hockey, and Pass-a-Puck in chapter 10.

APPENDIX C

ANSWERS TO TRUE OR FALSE REVIEW QUESTIONS

REMEMBER THAT A PARTIALLY FALSE QUESTION IS CONSIDERED FALSE.

Preface

1. True

2. False. Simplification is not an easy process. But this task was well worth the effort.

3. False. *Secrets to Success in Sport & Play* was written to assist both those who are seeking personal help and those who wish to help others. This could include coaches, parents, players, teachers, and anyone else who is interested in puzzling out the reasons why things happen in sports and play.

4. True

Chapter 1
Teaching and Learning

1. False. As the name implies (the specific objective method), you must first determine your objective—that is, what you want to accomplish. Which specific sport you want to deal with has usually been determined before you start your specific objective analysis.

2. False. Thumb through the book to find the orientation of this book.

3. True

4. True

5. True

6. False. I have found that learners tend to focus on the results rather than concentrate on the demonstration, so I no longer demonstrate by hitting a ball over the net or shooting a ball into the basket. Learners retain more of the relevant aspects of the demonstration when the result becomes irrelevant.

7. True

8. True. "The principles of balance, force production, motion and leverage are identical regardless of the activity" (Broer and Zernicke 1979, p. 29). While the mechanical principles do not change, it is the choices you make based on what you are trying to accomplish within a specific task that do change.

9. False. The basic foundation skills—such as absorbing force, spatial (space) awareness, and stopping and starting—remain the same across sports and play. Adaptations and applications depend on which of these are of greatest importance in relation to your specific objectives for the given situation. Because most actions are a result of a combination of the application of more than one mechanical principle, each can make a contribution commensurate with its value to your goals.

Chapter 2 Balance

1. True

2. False. Balance is often overlooked as an important component of skillful performance.

3. False. If we shift our weight (for instance, raise an arm, putting more body weight above the waist), the center of gravity must also shift in this direction to continue to balance the weight. Change the word *opposite* to *same* in the true–false statement and the answer becomes true.

4. True

5. True. Yes, bending the knees is a way to both lower the center of gravity and allow the center of gravity to be adjusted over the base of support. It does not increase the base of support, which is the third balance principle.

6. False. Be careful of the word *wide,* which indicates from side to side. The key is to increase the size of the base of support in the direction in which you may need to adjust your balance. A base of support that is wide but not in the direction of the oncoming force may actually reduce your adaptive ability to maintain your balance. Leaning into the oncoming force may not only allow you to absorb the force gradually in a rocking motion, but rocking backwards to absorb the force may also be useful as a preparatory movement for your next action.

7. False. Does a blocker lift the opponent's center of gravity? Can you see why pushing without lifting would not be as effective? Might a wrestler push and lift in a similar way?

8. True. Note the balance games and activities in chapter 10.

Chapter 3 Initiating Movement

1. False. Inertia is sometimes thought of as a state of stillness, but this definition is incomplete because inertia can be a still or moving state. Inertia is actually any present state of being, and overcoming inertia involves a change in that state and the resistance to that change. It could even involve a change in either speed (faster or slower) or direction.

2. True. Overcoming inertia will affect the amount of force that can be generated. Using the pull of gravity to overcome inertia will assist in initiating a task quickly. While the principles of overcoming inertia remain the same, how to best overcome inertia is based upon the task and what we are trying to accomplish by the task.

3. True. When you play, be sure you warm up and remain warm. This will allow you to move more quickly and forcefully and to avoid injuries that could occur if you were not sufficiently internally warmed. You save time if you already have your involved joints bent or unlocked and if the muscles to be used are on a partial stretch. Bending the joints in preparation for the move can put the muscles to be used on a partial stretch.

4. False. Because the question reads "partial stretch," it is false. For this to be a true statement, the word *full* would have to replace the word *partial.*

5. False. Players must not confuse the use of the stretch reflex to rapidly overcome inertia (thus allowing more force to be developed over the available distance) with the rapid completion of an action. They need to be aware that evoking the stretch reflex requires additional time during the preparation phase and thus would not be used in a situation that involves the need to rapidly complete a task. A short throw to first base, a flick in badminton, a block volley at the net in tennis, and a bunt in softball or baseball all require little force but need a rapid response for success.

6. True

7. True

8. False. Short levers do get you there faster, but they decrease force development. There are times when force may become secondary, less important, or even unnecessary. If a quick move or change is needed and if force is not a primary factor, then shortening the involved levers may help.

Chapter 4
Force

1. False. Opposition (involving opposite sides of the upper and lower body in an action) and reciprocal innervation (a process by which dual messages are sent to the muscles—specific muscles are stimulated to contract, while the opposing muscle groups receive a message to relax) are different. For more information about opposition, go to page 39; for information on reciprocal innervation, go to page 34.

2. True. The stretch reflex should not be involved if the primary objective is to complete the task quickly, and maximum force is not needed. The stretch reflex requires additional time and will increase the time to complete the task.

3. True

4. True

5. True. Absorption of force would occur if there was no stabilization.

6. False. The abdominal muscles stabilize the pelvic area and create the anchor point around which the movement and weight transfer take place in force development. Many less experienced participants have neglected the role that the abdominal stabilizers play in force production. Many weekend golfers, for instance, are not aware of the role that the abdominal muscles play in the distance of their drives (see figure 4.3). The abdominal muscles play a similar role in soccer, and the shoulder girdle has a stabilizing role in throwing and striking.

7. True

8. True

9. False. Flattening the swinging arc is extremely helpful to both the beginner and the advanced player. It allows the highly skilled player to strike or release later in the flattened arc pattern, thus increasing the time and distance over which force can be developed before contact or release. It allows the beginner, whose timing may be less than perfect, to strike the object or release the throw at several points in the forward movement without sacrificing accuracy because the arc faces the same direction throughout the flattened part of the swing.

10. True

11. False. This is true for all the factors *except* shortening the lever.

12. True. Some people wonder how the follow-through, which follows the application of force, can have an effect on force application. A complete and extended follow-through ensures that the slowing down process will not be initiated too early during the final part of the action phase. This means that maximum velocity is still available at impact, release, or other time of need. The final weight transfer continues over a bent forward knee into the follow-through. This allows for the gradual absorption of force.

13. False. This statement is false only because "spreading" the force is not done by bending the joints, but rather by increasing the area of impact.

14. True

15. True. There are some exceptions to the use of opposition in specific sports or for specific reasons. Some badminton players have preferred not to use opposition in the low, short serve in order to protect more easily against a quick backhand return. Fencers do not use opposition because they prefer to make minimal contact surface available to their opponent. Dart throwers do not use opposition in order to avoid body rotation that could reduce accuracy.

16. False. It is true that lever length affects both rotary speed and force; however, although a shortened lever can be brought around or rotated more rapidly, it does not create greater force. The lengthened lever may be slower and more difficult to rotate, but it will be able to build up more velocity over the additional distance. Thus, a full reach (not crowding the ball or not choking up on the striking instrument) may help a player hit a home run or have a more forceful tennis stroke, but it will require a longer time to execute. This creates no time problem in movements such as the tennis serve or a golf drive, but a fast pitch might get by you during the time required to take the "long" swing. When this occurs, you may choose a shorter lever with less force so that you can get around in time to meet the ball. If the oncoming object is traveling fast, you may be able to use the momentum of the ball to contribute to your force rather than develop force through the use of a long lever.

17. True

Chapter 5
Preparing Your Body to Apply Mechanical Principles

1. True

2. False. Forward shoulders could be caused by *weak* upper back muscles or *tight* chest muscles or both.

3. True

4. True

5. True. We need length of muscles for range of motion to increase the time and distance over which force can be developed. Without periodic stretching, this freedom and range of motion can be gradually diminished.

6. True

7. False. Weakness in the upper back and shoulder muscles could reduce the preparatory backswing in a throw or racket stroke, thus reducing force development. It is true that tight chest muscles can reduce a basketball player's reach for a rebound.

8. False. Connective tissue in muscles is common among physically active participants and is often insufficiently stretched in ways to prevent shortening and tightness. Without periodic stretching, freedom and range of motion can be gradually diminished even while the participant is still actively involved.

9. False. The use of ballistic or forceful bouncing to stretch a muscle is not the way to increase ROM (range of motion). Muscles can be lengthened, or their length can be maintained, by using periodic, SLOW stretching. A muscle that is or could become shortened should be brought to its full length and held on the stretch for several seconds. This should be done several times a week. Ballistic or forceful bounce stretching can cause minor muscle tears and the involvement of the stretch reflex, both of which can reduce the desired effect of lengthening the stretched muscle.

10. True. Tight quadriceps limit the backswing of the leg, reducing the time and distance over which force can be developed. Tight hamstrings can limit the forward swing of the leg, also reducing the time and distance over which force can be developed.

Chapter 6
Rebound, Deflection, Spins, and Roll Patterns

1. True

2. False. A bowling ball has an optimum (best) speed that is most effective. If it is traveling too fast, it could actually reduce the amount of "mixing" and thus the number of pins knocked down.

3. True

4. True

5. False. A ball with backspin gives an additional push forward against the contacted surface. The additional reaction is backward, but as long as the forward rebound force is greater than the backward spin force, the ball will still move forward. The resultant force will cause the ball to rebound higher (the attempt to go backward), and because of the two forces moving in opposite directions, the ball will slow down and not go as far.

6. False. The topspin on a foul shot that hits the backboard will push the ball *upward* away from the basket. On the other hand, if a foul shot that hits the backboard has backspin, this spin will push the ball downward toward the basket, and the friction caused by the spin may slow the ball down, allowing gravity to also pull the ball down toward the basket.

7. True

8. True

9. False. The important thing is which way the hook ball spin will be pushing against the floor. Remember that for every action (in this case, push) there is an equal and opposite reaction. For a left-handed hook, the bottom of the ball will be pushing left, causing the ball to travel to the *right* when the ball slows down. This allows it to hook into the kingpin. The spin of the ball and the deflection of the 5 pin against other pins make a well-controlled hook ball very effective.

10. True

Chapter 7
Projectiles

1. True. The flight pattern of an object is normally fairly consistent. Unless modified by wind, spin, or some characteristic of the object that affects air pressure in a specific way, the pattern is parabolic. This simply means that the second half of the flight mirrors, or duplicates, the first half of the pattern in reverse. Being aware of this may help those sending and receiving an object to determine the path it will take and where it will come down. Thus, they become more able to accurately direct a flight and to read the flight of an object.

2. True

3. True

4. False. The first part of this statement ("When the body leaves the ground and becomes a projectile, the center of gravity follows a parabolic path") is true. The second half of this statement ("but we can change the parabolic path of the center of gravity by changing the position of various body parts") is false. Although you can change the position of the center of gravity within your body during flight by changing the position of your body parts, you cannot change the flight path of the center of gravity once your body is airborne. So, your body position in the air will change in relation to the new weight distribution, while the center of gravity of the body retains its original parabolic flight pattern. This is a rather difficult concept, but once understood, it can be quite helpful in making decisions.

5. True. If you reposition your weight in any direction by moving a body part (i.e., moving arms, legs, or head or bending the trunk), your center of gravity will also shift in that direction within the body to divide the newly distributed weight around the center of gravity. For instance, if you lift your leg to the right, your center of gravity will shift up and to the right within your body to become your new balanced center of weight distribution.

6. False. The first part of the statement ("When the arms are raised, the center of gravity within the body is raised") is true. But the second part ("as is the body position in the air") is not true. The body will actually be lower in the air. If your center of gravity shifts upward in your body because you raise your arms, but the path of your center of gravity cannot change, then the position of your feet in the air will be lower than before you raised your arms. The opposite also holds true. If you lower your arms during flight, the position of your feet should be higher than if you did not lower your arms. Timing will be important in using this factor effectively. This is another difficult concept and may require studying pages 68 to 69.

7. True. Considering that the parabolic path of the center of gravity does not change once the body is in flight, a person could put his feet farther *forward* in a broad jump by having his arms back just *before* landing (causing the person's center of gravity to be farther backward in his body and causing his legs to be farther forward than if both his arms and legs were forward). The person would still need to bring his arms forward on landing in order to bring his center of gravity forward so that he would not lose balance and fall backward. Exceptional timing skill would be required to use this effectively.

8. False. The high jump technique was not modified by changing the path of the center of gravity, but rather by changing the position of the center of gravity within the body. A skillful individual is now able to go over the high jump bar at a higher level by moving the center of gravity within or perhaps outside of the body, thus allowing the body to be positioned to avoid the bar.

9. True. The technique used in each of the high jumps discussed becomes progressively more sophisticated, and for participants without skill, the risk of injury also increases. For this reason, beginners should start with the simpler techniques.

10. True. Lowering a body part will lower the center of gravity within the body, thus raising the body in the air as the path of the center of gravity stays the same (once the body has become airborne). This allows the hurdler to clear the hurdle, while maintaining a lower body position. Bringing the lead leg down quickly to take the next step does the same thing.

Chapter 8
Direction and Accuracy

1. True

2. False. Other forms of air resistance besides spin can be used to a player's advantage. Track and field events and the kick in football are two examples where wind blowing in an advantageous direction can benefit the participants.

3. False. In this book, the term *flattening the swinging arc* refers to the path (arc) made by the body part (hand or foot) or striking implement (bat, racket, or club) as it swings around the body. This is not the same as swinging level, which normally refers to swinging in the same plane in relation to the ground.

4. True. Flattening the swinging arc allows for some flexibility or possible errors in the timing of the swing. This is extremely helpful to the beginner whose timing (and thus accuracy) may be less than perfect. Flattening the swinging arc also allows the highly skilled player to strike or release later in the flattened arc, which makes it possible to increase the time and distance over which additional force can be developed.

5. False. See the information provided in the answer to question 4.

6. True

7. True

8. False. You do not want slowing down or absorption of force during any part of the action phase. A complete and extended follow-through ensures that the slowing down process will not be initiated too early during the final part of the action phase.

9. True. If you want to run, walk, or swim forward, then whenever possible all your push or pull should be in a backward direction. Any push or pull not directly in line with and opposite to the desired reaction can lead to less desirable results. The action–reaction principle also relates to angles of rebound, contact points, direction of impact force, flattening of the swinging arc, and timing of the release or striking point.

10. True

11. True

12. False. By studying resultant force on page 82 and spins on pages 58 to 61, you can gain a more thorough understanding of why you would use a different spin from each side of the basket.

Chapter 9
Visual Evaluation

1. True

2. False. Both being aware of the basic principles of movement and participating in carefully selected and related activities can help reduce the time required to gain and improve skills.

3. True

4. True

5. False. Skills are developmental, and our task is to find what aspect or aspects need help and how we can creatively and effectively extend the needed help. A player's optimal ability may never be reached if the foundation skills are left incompletely developed or are neglected because the player can hit or catch a ball and is having some success. We need to find enjoyable activities to give each participant challenging experiences that allow for continuous growth and development.

Chapter 10
Activities to Habituate the Mechanical Principles of Movement

1. False. The first sentence is true, but the second sentence is not. The real function of the games in chapter 10 is to draw responses from the players that help them to discover seven common mechanical principles of movement. These principles support successful play and sport techniques. This approach will improve players' playing abilities by helping them to intuit and then apply these mechanical principles of movement.

2. True

3. False. Although coaches, teachers, and parents should understand the mechanical principles of movement, it is not their job to explain these to the players. This approach is called nonverbal guided discovery through play, or learning without words. As the players discover the mechanical principles through the play experience, these can be intuited and applied by the players in many play and sport activities.

4. True. "The purpose of each (movement pattern) causes some adjustments, but the basic mechanics remain the same" (Broer and Zernicke 1979). The goal of a given task (e.g., Do we need maximum force for a home run or do we need to minimize force and move quickly for a bunt?) determines how we choose to apply various aspects of the related principles.

5. False. Flattening of the swinging arc can lead to greater force and accuracy in many sport skills. Participants can transfer the components involved in the flattening of the arc into many sports; however, diving deals with the parabolic path of the center of gravity (page 68) and the length of the lever in relation to the speed of rotation (pages 40-41), but not the flattening of the swinging arc.

6. True

7. True

8. True

9. True

10. True

11. True

12. False. The reverse is actually true. Pass-a-Puck is a good lead up to Rag or Rug Hockey.

13. False. All four mechanical principles of movement (contact points, flattening the swinging arc, angle of the striking surface, and follow-through) will transfer to tennis.

14. False. The team that is two points *behind* gets to choose a member of the opposing team to become a twister.

15. False. Putting players on these skates allows you to have more players participating vigorously in less space with reduced potential for injuries if collisions occur.

16. False. This statement is incomplete and incorrect when it says, "In Rebound Board, players are practicing *only* the important mechanical principles related to contact points and angles of the striking surface." Playing with the rebound board also enables players to practice the mechanical principles of flattening the swinging arc, following through, and reading the parabolic path.

17. True

18. True

19. False. Playing on skates does develop the abdominal muscles that control the pelvic region, which adjusts the center of gravity over the base of support to maintain balance; however, the abdominal muscles also stabilize the hips for greater force production in sports such as golf, soccer, tennis, and gymnastics.

20. False. Super Sox supports both balance and absorption of force, but it has little to do with follow-through. A reader might interpret "follow-through" in Super Sox as the push of the socks onto the arms of the receiver. If this happens, the answer to this true–false statement could be debated.

21. True. The activity of Super Sox tends to force players to take a natural force-absorbing stance (forward and backward stride) and bend the knees to absorb force as the socks are pushed vigorously onto their arms. This is the same stance that a person would take to effectively absorb and successfully receive any oncoming force. By becoming habituated to these force-absorbing techniques, players can transfer and apply the techniques in other situations.

INDEX

Note: The italicized *f* and *t* following page numbers refer to figures and tables, respectively. For readers looking for a more detailed index, the author has created a custom index available for free at www.humankinetics.com/products/all-products/Secrets-to-Success-in-Sport--Play---2nd-Edition. Space did not allow for both indexes in this book.

ABOUT THE AUTHOR

Marianne Torbert, PhD, is director of the Leonard Gordon Institute for Human Development Through Play of Temple University in Philadelphia. A street athlete at age 6 and a national participant at 16, Marianne Torbert has played, studied play, and taught kinesiology for several years. She holds a PhD from the University of Southern California and is a professor emeritus at Temple University. She is also the author of *Follow Me: A Handbook of Movement Activities for Children.*